Oil Property Evaluation

Second Edition

Robert S. Thompson

Colorado School of Mines

John D. Wright

Colorado School of Mines

P.O. Box 892
Golden, Colorado 80402

1985.01

PREFACE

Oil Property Evaluation is an attempt to provide a practical, useful, and consistent text for people directly involved with evaluating oil properties on a day to day basis.

Since we view it as a first text in oil property evaluation, it covers some basic concepts in an effort to alleviate much of the confusion arising from the (mis) use of accounting terminology in the evaluation of projects. From there, the subjects of time value of money and measures of investment worth are discussed. This ends the "generic" portion of the book. The remainder is devoted exclusively to the evaluation of oil and gas properties and investments. The subjects covered include reserve analysis, production forecasting, oil and gas pricing, costs, determining working and net revenue interest, severance and advalorem taxes, calculating "windfall profit" and federal income taxes, and a short discussion of the evaluation report.

It is our intention that, with the reader supplying current prices and costs, this text can serve as the sole reference necessary to evaluate an oil property. In order to accomplish this, we have been forced to include some information which is dated. The sections on taxes (severance and advolorem, windfall profits, and federal income) are subject to change at any moment at the whim of various legislatures. We have presented the most up-to-date information that we have and will continue to modify and edit the book as necessary to reflect the latest changes.

Aside from this material, the remainder of the book presents concepts and methods which are applicable throughout time and throughout the world with little modification. Of particular interest is the detailed cash flow diagram in Chapter 1. This diagram is independent of the tax laws in effect at a given time and will work for many foreign countries as well as the United States.

Some of the material, such as the method of calculating working and net revenue interests, and the flow diagram for income taxation, is unique and original. Much of the rest is a reorganization and re-statement of concepts the authors have learned from the many people they have studied under and worked with as well those countless authors who have put their ideas in writing. The authors gratefully acknowledge their contributions.

We would like to thank Sharon Dehmlow for her fine job of typing and Karen Carmichael for proofreading and expediting. A special thank you is in order for our wives, Geri Thompson and Jane Wright, without whose continued support this project would never have been completed.

It is our intent to revise and expand this text periodically. To aid us in that effort, we would appreciate your comments and suggestions.

Robert S. Thompson
John D. Wright

Golden, Colorado
July, 1984

ABOUT THE AUTHORS

Robert S. Thompson graduated from Colorado School of Mines in 1969 with a professional degree in petroleum engineering and received his M.B.A. degree from the University of Houston in 1977. Presently, Robert is an assistant professor at Colorado School of Mines teaching courses in oil property evaluation, production engineering, and hydrocarbon fluid behavior. He has worked as a petroleum consultant performing oil property evaluations and reservoir simulations. Other industry experience includes both onshore and offshore petroleum production operations, systems design, and corporate financial planning. While working for a major oil company, he was responsible for corporate exploration and production long range plans, capital budgets, and strategic studies.

Robert is a registered professional engineer and a member of the Society of Petroleum Engineers.

John D. Wright graduated from Colorado School of Mines in 1969 with a professional degree in petroleum engineering. He worked for Conoco in Ventura and New Orleans as a production engineer. After returning to graduate school at CSM, he taught for a year and then, in 1975, co-founded Jerry R. Bergeson and Associates, an international consulting firm. While at Bergeson and Associates, he performed numerous property evaluations for bank loans and SEC filings. John also made engineering studies on reservoirs in the United States, Canada, Europe, and the North Sea, in addition to completing wells in a number of western states. In conjunction with this work, he wrote a number of computer programs, including a commercially available oil property evaluation package.

In 1980, John returned to CSM as a full-time instructor in petroleum engineering while continuing work on a Ph.D. He has taught undergraduate courses in oil property evaluation, production, and rock properties and graduate classes in petroleum economics and production.

John is a registered professional engineer and a member of SPE-AIME and SPWLA. He has served as treasurer of the Denver Section of SPE.

The book **Oil Property Evaluation** is currently being used as a text in numerous universities. The book is also used as a text in a short course, Applied Oil Property Evaluation, taught by the authors.

Oil Property Evaluation

Contents

Detailed table of contents are presented at the beginning of each chapter

---

Chapter 1: Basic Concepts

Chapter 2: Time Value of Money

Chapter 3: Investment Decision Analysis

Chapter 4: Reserve Estimation

Chapter 5: Decline Curves

Chapter 6: Oil and Gas Prices

Chapter 7: Oil Field Deals

Chapter 8: Costs

Chapter 9: Severance and Advalorem Taxes

Chapter 10: Windfall Profit Tax

Chapter 11: Federal Income Taxation of Oil and Gas Transactions

Chapter 12: The Report

Appendices:

A: Derivation of Geometric Gradient Formula

B: The Use of Loss Ratio to Extrapolate Decline Curves

C: Construction of Specially Scaled Graph Paper

D: Depreciation Methods Prior to 1981

Index

Table of Contents
Chapter 1
Basic Concepts

1.1 Profit 1-1
1.2 Cash Flow Model 1-2
1.3 Financial Model 1-4
1.4 Tax Model 1-9
1.5 Which Model is Best? 1-10
1.6 Incremental Analysis 1-11
1.7 Uncertainty (a Form of Risk) 1-13
1.8 Cash Flow Diagram for an Oil Property 1-13

Chapter 1

Basic Concepts

An oil property may be evaluated for many reasons. In many cases the reason behind the evaluation dictates the methodology employed and the form of the results. For example, an evaluation prepared for the Securities and Exchange Commission (SEC) will use constant prices and costs while an evaluation to obtain a bank loan will normally use escalated prices and costs. A few of the reasons for performing an evaluation are listed below:

- Purchase of Producing Property
- Sale of Producing Property
- Collateralization of Bank Loans
- Basis for Decision to Invest in Project
- Choosing Among Alternatives
- Governmental Reporting
  - SEC
  - Department of Energy

In many cases, the evaluation will form the basis for a decision (such as drill or don't drill). In these cases the report must address the goals of the organization. Usually the primary goal is to maximize present value profit, but many other goals are possible. Some corporations are concerned about retaining a AA bond rating. Others are concerned about longevity. Still others may be more concerned with avoiding risk than with maximizing present value profits. The engineer must keep the goals of the organization in mind when he performs an evaluation.

In this text, we will primarily be concerned with maximizing present day profit. This should not imply, however, that this is the only legitimate goal—it is just easier to quantify.

1.1 PROFIT

Having stated our primary goal in terms of "profit," it is now time to ask, "What is profit?" This seemingly trivial question is the root of a tremendous amount of the confusion that is present in the evaluation field today. The small businessman, the accountant, and the tax attorney all talk about profit, but usually they are all referring to totally different things. To top it all off, they all use the same (or similar) terminology.

Let us define profit by the following equation:

Profit = Revenues — Costs

This seems simple enough. Usually (but not always) we can agree on the revenues. Revenue is the gross amount of money received for our services or products. If we sell 100 barrels of oil at $32.00 per barrel our revenues are $3,200.00. The only real problem is determining *when* the revenue is to be "recognized." This problem is minor, however, when compared to the problem of defining costs.

There are three common "models" which are used to determine profit. Each model (cash flow, financial, and tax) has a different way of defining costs, and therefore, results in a different calculated value of profit. Before we compare the three models, let's discuss one more concept—*periodicity*.

Webster defines periodicity as "the state, quality, or fact of recurring at regular intervals." Profit has periodicity—it keeps happening at regular intervals (we hope). Profit is measured for an assumed time period. In an oil property evaluation the engineer must project profit for each of many future periods. Normally, the period size is one year. We could, however, estimate profit on a monthly basis or a daily basis or even on an hourly basis. When picking the time period and model to use, an effort should be made to maximize the benefit to cost ratio resulting from using that particular model and forecasting period. In other words, the incremental cost of using a more sophisticated model or forecasting technique may exceed the incremental benefits to be derived from using that model or technique. The engineer should also take care not to confuse precision and accuracy. Forecasting the revenue and costs for an oil property on a daily basis will allow us to be very *precise* in our estimate of when cash is received and paid out, but it is probably no more *accurate* than forecasting on a monthly or yearly basis.

The cash flow model has one unique feature. This is the zero length period referred to as ''Time 0''. Time 0 usually refers to the instant (day) when the first significant investment is going to be made. This might be the day the check for an offshore lease will be written or the day when the check for the drilling of an onshore wildcat well will be written (assuming we already have the lease). Time 0 is the point to which all future profits are discounted. It is often referred to as the effective date of the study.

Now let's talk about the three profit models—cash, financial, and tax.

1.2 CASH FLOW MODEL

In the cash flow model the ''profit'' we have been referring to is more properly called Net Cash Flow (NCF). This highly descriptive term is attached to the net flow of cash, which has been generated by the project, into or out of the treasury. In the cash flow model, revenues are ''recognized'' when the cash is received and costs are recognized when the cash is paid out. A flow chart for calculating net cash flow after Federal income taxes (AFIT) is shown below. This flow chart can also be used to calculate the net cash flow before Federal income taxes (BFIT) by simply omitting the tax calculation. The bottom line then becomes net cash flow BFIT.

Your share of Gross Revenues
– Your share of Direct Costs
= Net Operating Income BFIT
– Income Taxes
= Net Operating Income AFIT
– Investments
= Net Cash Flow AFIT

For example, if, on June 30, 1985, an oil company intends to write a check to a drilling contractor for $1,000,000 to drill an oil well, then that day becomes Time 0 for the evaluation of the well. The profit for the project *on that day* is $-1,000,000. Assuming that the well is successful, the estimated profit for each year will be the cash received from oil sales less the direct expenses, income taxes and investments in that year. Figure 1-1 presents a cash flow diagram. A cash flow diagram is a technique that is helpful in analyzing profit when profit is based on net cash flow. Notice the profit for the time 0 period is represented as a down arrow, thus indicating that the profit for that instant is negative (the $1,000,000 paid to the drilling contractor). Net cash flow is then measured for each subsequent assumed time period. In Figure 1-1, the net cash flow (profit) for each period is assumed to be received at the end of each period. In using the cash flow model to measure profit, care should be taken to include only cash items. An investment (negative NCF) that burdens the profit for the time 0 period should not burden the profit in any subsequent time period. Remember, only cash transactions are used to calculate profit.

In order to show the difference between the three models, the ''profit'' calculated by each model for an investment in a hypothetical oil well will be presented in this chapter. The profit calculated by the cash flow model is shown as Example 1-1.

Example 1-1
Using the Cash Flow Model to Determine "Profit"

We are considering investing $60,000 in an oil well. For that amount of money, we are able to purchase only a portion of the total well, but that is unimportant at this point. Our analysis will involve only our investment and our share of revenues and costs. We have projected the following revenues, costs, taxes, and net cash flows for our interest in the well.

PROJECTIONS

Year	Revenues $	Direct Costs $	Income Taxes $	Investment $	"Profit" Net Cash Flow $
0	—	—	—	60,000	—60,000
1	58,800	21,900	—200	0	37,100
2	45,100	19,300	9,000	0	16,800
3	34,700	17,800	4,700	0	12,200
4	26,600	15,300	2,660	0	8,640
5	20,400	15,500	—540	0	5,440
6	7,000	6,500	250	0	250
	192,600	96,300	15,870	60,000	20,430

Notice that the proposed investment of $60,000 occurs at Time 0. The "direct costs" column includes projected operating costs, severance and advalorem taxes, and the windfall profits tax. Details of the methods used to calculate the direct costs and income taxes are covered in later chapters.

FIGURE 1-1
Cash Flow Diagram

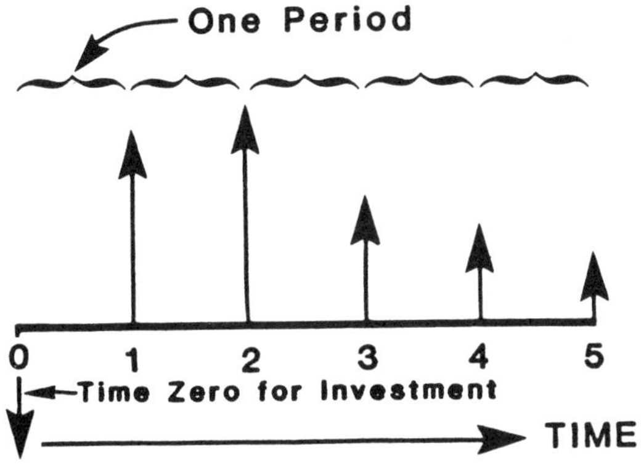

A few comments are in order. First, notice that every number shown in the example is a "real" dollar number. That is, every item presented in the example is a cash figure. Profit, or net cash flow, is calculated by assuming that revenue is cash in and that direct costs, income taxes, and the investment are all cash out. Note the negative income tax in Year 1 and Year 5. No, it doesn't mean that the government writes you a check (usually). It means that, taken as a whole, your company will pay less taxes in that year if it does the project than it will pay if it doesn't do the project. In this case, since your project generates the tax savings, the project should get credit for it. A more formal method of presenting this concept appears later in this chapter.

What happens if the tax savings from a project is greater than the company's ordinary cash tax liability for a given year? Depending upon the taxpayer's circumstances, the tax savings may be carried back to offset taxes in prior years or possibly carried forward to offset taxes in subsequent years. This will be discussed in the chapter on income tax calculations.

1.3 FINANCIAL MODEL

In a financial model, recognition of revenues and expenses for the determination of financial net income can be on a cash basis, modified cash basis, or an accrual basis. Small firms typically use a modified cash basis. Large firms usually use the accrual method of accounting. In strict cash basis accounting, expenditures for capital assets such as plant and equipment would be expensed in the period in which the assets were purchased. In the modified cash basis and the accrual method of accounting, plant and equipment purchases (assets providing future benefits) would be capitalized and expensed in the current and future periods. ''Capitalize the cost'' is a phrase used to mean that the expenditure is for an item such as land or equipment that can benefit (help generate revenue in) future periods. Thus, when the cost is capitalized, the item becomes an asset to the firm. An expense, on the other hand, is the cost of an asset that is assumed to expire during an accounting period in an attempt to generate revenues during the period. Some items, such as direct labor and power costs, are expenses since expenditures for these items will not benefit any future periods. The portion of previously capitalized costs that are assumed to be ''used up'' during the current period in an effort to generate revenues is also considered an expense. It is the accountant's attempt to match revenues and expenses for an accounting period that introduced accrual accounting and the terminology *depreciation, depletion,* and *amortization.* Since many sales and purchases are made on account, that is, credit is either given to a customer or supplied by a vendor, the flow of cash often lags the transaction recognizing the revenue or expense. An accrual, therefore, is when the recognition of revenue or an expense precedes the cash transaction. An example of the difference between a cash transaction and accrual accounting is the handling of crude oil sales. If our year end is December 31, 1983, for example, the accrual method would credit 1983 for crude oil pipelined or trucked during December even though the cash had not been received from the purchaser. The cash method of accounting, however, would credit 1984 since the cash wouldn't be received until January (or February or ...). For oil properties, the revenue streams for the cash and accrual methods are usually not materially different. Accordingly, the assumption of accrual accounting for revenues is not particularly a problem. Direct operating costs are usually handled in a manner similar to revenues—they are ''expensed'' or charged as a cost in the period in which the liability is *incurred*—not necessarily when it is paid. That is, the flow of cash may lag the transaction recognizing the expense. Again, for oil properties, the difference between cash and accrual accounting is probably immaterial when accounting for revenues and direct costs. Figures 1-2 and 1-3 demonstrate the accrual concept for revenues and direct costs.

Where the financial model differs radically from the cash flow model is in the handling of capital costs (those costs which will produce revenue in future periods as well as the current one). One of the principles of financial accounting is the matching principle. This principle states that the revenue from a period should be burdened with the cost of producing that revenue. Furthermore, the cost of producing revenue in future periods should not burden revenues from the current period. In the case of our oil well, the financial accountant would argue that, since the well generates income for a number of years, the cost of drilling and equipping the well should be spread out over the life of the well.

Three methods of systematically expensing the cost of a capital asset over the assumed life of the asset are called *depreciation, depletion,* and *amortization*. These terms will be discussed in the con-

FIGURE 1-2

Flow of Revenue in Financial Accounting

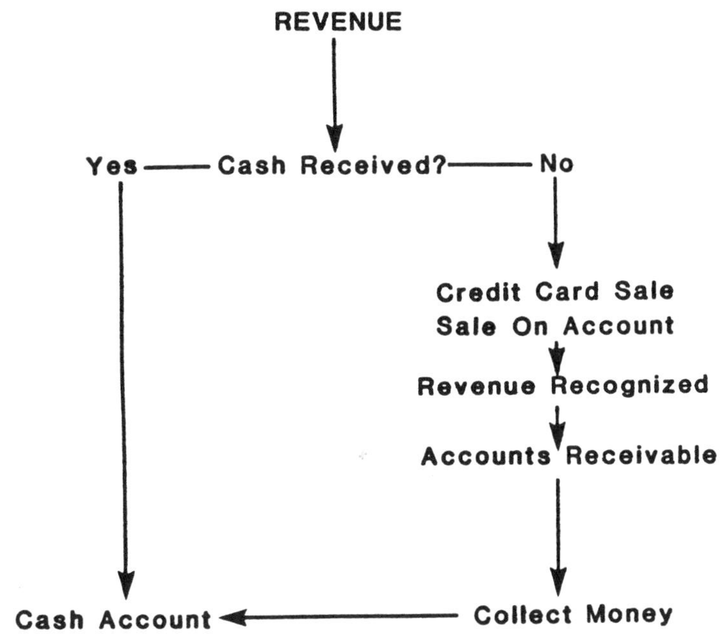

FIGURE 1-3

Flow of Operating Expenses in Financial Accounting

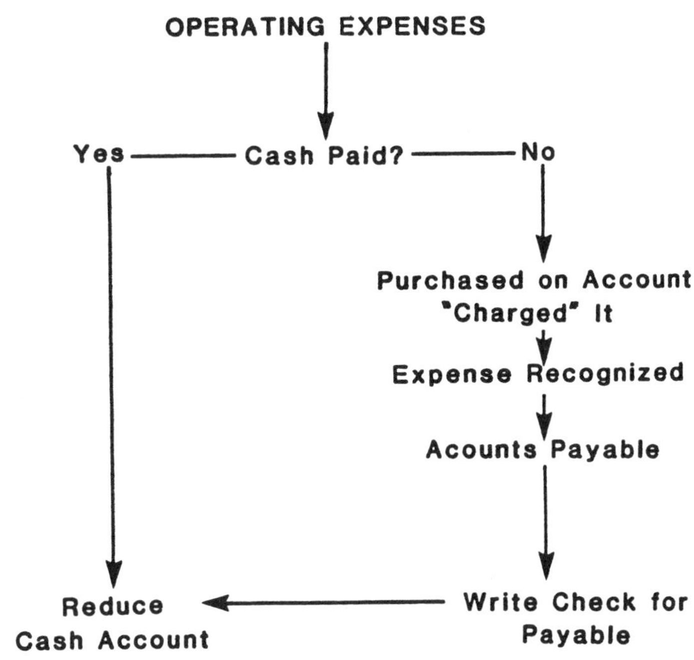

text of *Financial Accounting* (as opposed to income taxation).

Depreciation is the systematic expensing of the cost of a tangible asset over the assumed life of the asset. Note the terms systematic and assumed life in the definition. The financial accountant burdens future periods with capital expenditures in a systematic manner. This systematic method may not reflect a true matching of revenues and expenses. Additionally, the assumed life is not necessarily indicative of the actual producing life of the asset. Examples of tangible assets include buildings, cars, pumping units, and tanks.

Although other methods exist, by far the most popular methods are "straight-line" depreciation and "units-of-production" depreciation. In straight-line depreciation, the capital cost is allocated to future periods based on the assumed life. If the $60,000 oil well in our example is expected to last 6 years, then the revenue in each of those 6 years is charged with $60,000/6 or $10,000. Since oil wells normally produce less and less revenue each year, it can be argued that the straight-line method burdens early revenues too little and late revenues too much. Units-of-production (UOP) depreciation alleviates this problem. The cost of the well is charged against revenue based on the fraction of ultimate reserves produced each period. If 20% of the ultimate reserves are produced in the first year and 17% are produced in the second year then $12,000 would be written off in the first year and $10,200 would be written off in the second year.

Depletion is the systematic expensing of the asset value of a natural resource over the assumed life of the asset. A lease bonus or the purchase price of an oil property in excess of the value of the tangible equipment are examples of depletable items. Depletion is calculated in the same manner as units of production depreciation. The method of calculating the yearly charge is the same—the difference is that depletion is applied to the cost of acquiring property (lease bonus or purchase price) and depreciation is applied to tangible equipment (casing, tubing, tanks, etc.).

Amortization is the systematic expensing of an intangible asset (typically goodwill) over the assumed life of the asset. It is normally calculated using the straight-line method or the units of production method.

Collectively depreciation, depletion, and amortization are referred to as DD&A. They are also referred to as "non-cash charges" since they do not require the outlay of any cash. Figure 1-4 demonstrates the concept of capitalize versus expense and the use of DD&A.

Financial Accounting with DD&A results in a totally different "profit" picture than the cash model. The "profit" from the financial model is more properly called "financial net income."

A flow chart for calculating financial net income would be as follows:

  Your share of Gross Revenues
– Your share of Direct Costs
– DD&A
– Income Taxes
= Financial Net Income AFIT

The financial net income for our oil well is shown in Example 1-2. By assuming that constant prices were used, the proportion of revenue received each year is the same as the proportion of product sold. This allows UOP depreciation to be applied to the $60,000 investment.

Notice that the only difference between the financial and cash flow models is in their handling of the capital cost. In the cash flow model the $60,000 is "expensed" at Time 0. In the financial model, the cost is spread out over the life of the project. Notice that the total profit over the project life is the same for both models. Only the timing is different! Figure 1-5 compares the treatment of capital expenditures in the cash flow model with the treatment used in the financial model.

FIGURE 1-4

Flow of Expenditures In Financial Accounting

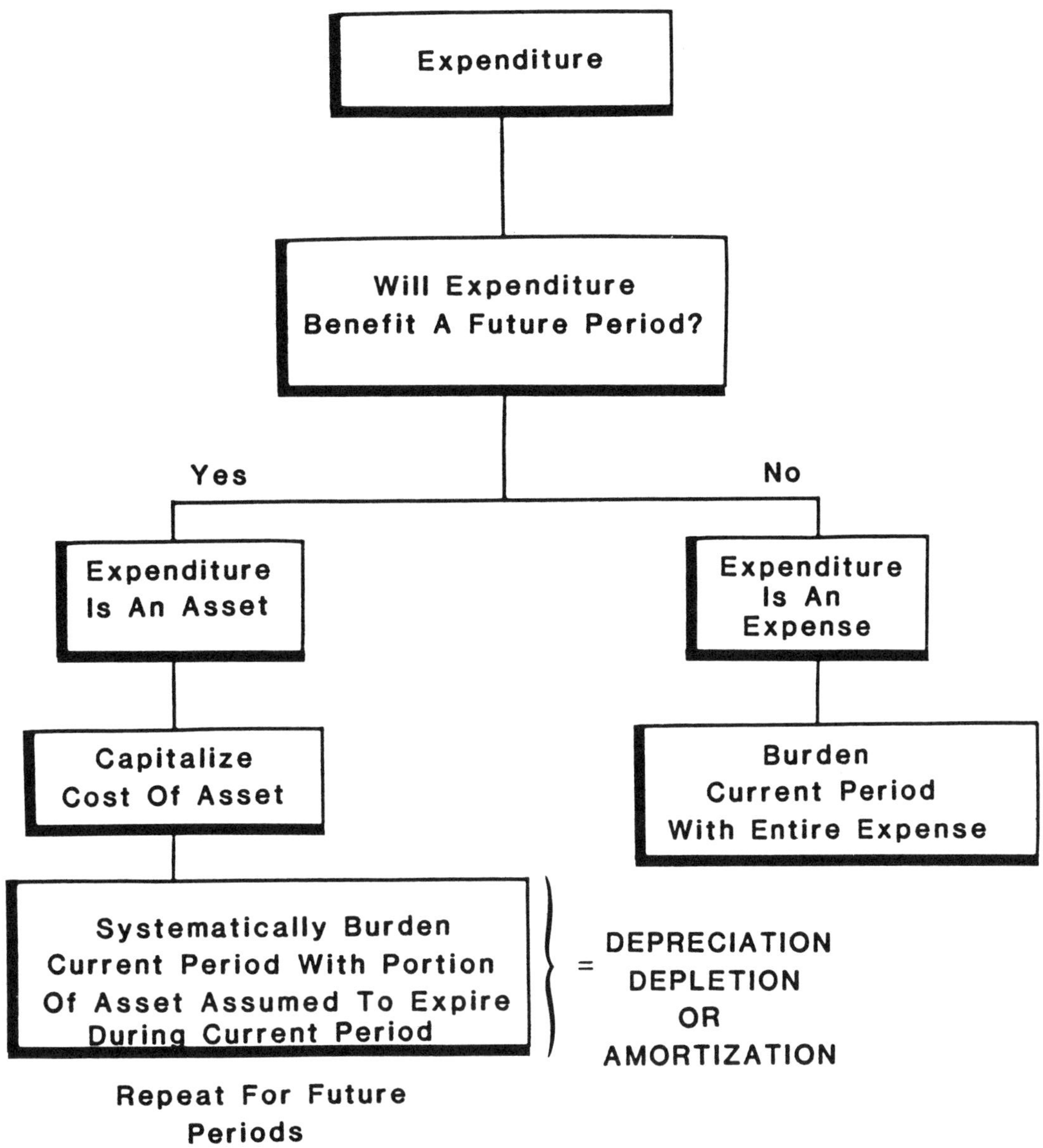

FIGURE 1-5
Capital Expenditures

CASH FLOW MODEL VS. FINANCIAL MODEL

Cash Flow Model

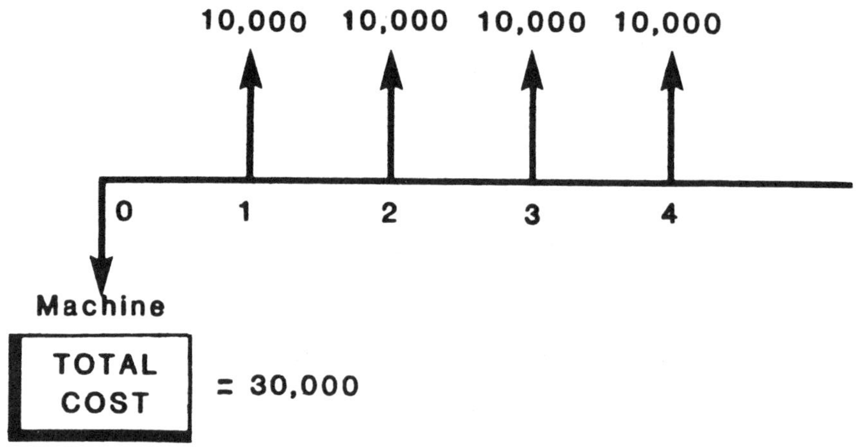

Financial Model

Assume Straight Line Depreciation and Four Year Life for Capital Expenditure

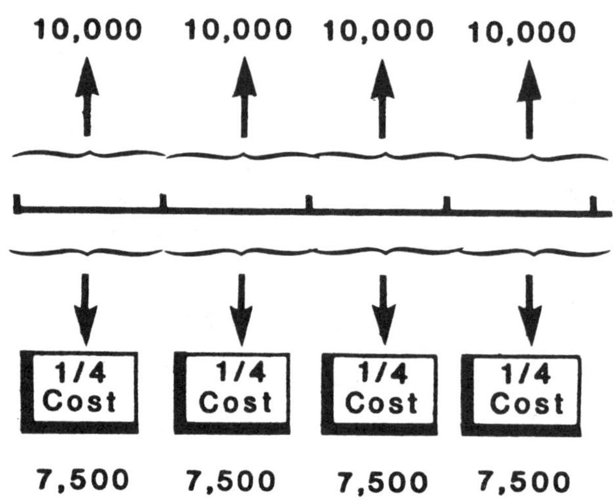

Example 1-2
Using the Financial Model to determine Profit
Use the data from Example 1-1
UOP depreciation for all $60,000
Constant Prices

PROJECTIONS

Year	Revenues $	Direct Costs $	Deprec-iation $	Income Taxes* $	"Profit" Financial Net Inc $
1	58,800	21,900	18,310	— 200	18,790
2	45,100	19,300	14,050	9,000	2,750
3	34,700	17,800	10,810	4,700	1,390
4	26,600	15,300	8,290	2,660	350
5	20,400	15,500	6,360	— 540	— 920
6	7,000	6,500	2,180	250	— 1930
	192,600	96,300	60,000	15,870	20,430

* Income Tax for the Financial Model is assumed to be the same as Income Tax for the Cash Flow Model. In the real world this is not true, but we are not going to discuss the finer points of accounting here.

1.4 TAX MODEL

The Internal Revenue Service (IRS) sets (and then changes) the rules for the calculation of "profit" in this model. The IRS recognizes that a portion of the revenue from a property represents the return of the invested capital and that a portion of the revenue represents rents or income generated by that capital. Under the economic system currently in use in the United States, it is desired that the portion which represents income be taxed and the portion which represents the return of capital not be taxed. Deciding how much of the revenue stream in a given year is return of capital and how much is taxable income or profit becomes quite complicated. Basically, "profit" is defined as revenues less the sum of direct costs, intangible drilling costs, depreciation, depletion, and amortization. Although the terms "DD&A" are also used in the financial model, DD&A in the tax model are calculated in a different manner. They are completely different numbers. For instance, depreciation is applied to tangible equipment—but not necessarily the same "tangibles" as used in financial accounting. The **method** of calculating depreciation is also different. Currently, under the tax laws of 1981 and 1982, a tangible item is depreciated using the "Accelerated Cost Recovery System" (ACRS). ACRS is a method of calculating depreciation based on a combination of accelerated (declining-balance and sum-of-the-years-digits) and straight-line methods. The IRS has published tables for calculating ACRS depreciation. ACRS was written into the tax law for the first time in 1981 and was changed in the 1982 tax law. The 1984 law may change it again. Details on the various methods of calculating depreciation will be presented in the chapter on income taxation.

Depletion, as used in the tax model, is also different from that used in the financial model—same name but different numbers.

The ''profit'' calculated in the tax model is more commonly called ''taxable income.'' A flow chart for calculating taxable income would be as follows:

Your share of Gross Revenues
— Your share of Direct Costs
— Allocated Overhead
— Expensed Intangible Drilling Costs
— DD&A
= Taxable Income

The tax model for our oil well is shown in Example 1-3 along with the assumptions used to generate taxable income. Further discussion of the calculations will be deferred to Chapter 11. Notice that both the yearly ''profit'' and total ''profit'' are totally different from that calculated by the other two models.

1.5 WHICH MODEL IS BEST?

All of these models—cash flow, financial, and tax— will be discussed at length in later chapters. They are presented here so that the reader will realize that there are at least three models in use with much of the same terminology. Which one should be used? For economic analysis the answer is simple— *always use the cash flow model.* The reason lies in the time value of money concept. As explained in the next chapter, money has a time value. One dollar received today has more value than one dollar received 5 years from now. This means that assigning the right ''profit'' to the right

Example 1-3
Simplified Tax Model of ''Profit''
Use the data from Example 1-1
$30,000 Tangible Equipment
$30,000 IDC
Major Oil Company

PROJECTIONS

Yr	Revenues $	Direct Costs $	Expensed IDC $	Amortized IDC $	Deprec-iation $	''Profit'' Taxable Income $
1	58,800	21,900	25,500	1,500	4,275	5,625
2	45,100	19,300	0	1,500	6,270	18,030
3	34,700	17,800	0	1,500	5,985	9,415
4	26,600	15,300	0	0	5,985	5,315
5	20,400	15,500	0	0	5,985	−1,085
6	7,000	6,500	0	0	0	500
	192,600	96,300	25,500	4,500	28,500	37,800

Income Tax for year 1 is .5*($5,625) — $3,000 (Investment Tax Credit)

Income Tax for all other years is .5*(Taxable Income)

period is very important in economic analysis. In fact, in some cases it is critical to making the correct decision. Since, in making an investment, we must lay out cash and cash is returned to us by the project, the cash flow model is the correct one to use. The other models distort the timing of the flow of funds to and from the project. In the case of financial net income and taxable income, the distortion is the result of arbitrary decisions made by the accountant or the IRS. Of course, when using the cash flow model for an after tax evaluation, the tax model should be used to calculate taxable income and cash income taxes.

The following table summarizes the meaning of profit for each of the three models.

MODEL	Better Name for 'Profit'
Cash	Net Cash Flow
Financial	Financial Net Income
Tax	Taxable Income

1.6 INCREMENTAL ANALYSIS

Before proceeding further, one of the concepts that should be understood is the concept of incremental analysis. This concept is often applied when comparing two mutually exclusive projects, but few people realize that it should also be applied when evaluating a single project. In order to show the true effect of doing a project, the projected cash flows for the company *without* the project should be subtracted from the projected cash flows of the company *with* the project. This sounds rather impractical when a billion dollar company is evaluating a $60,000 investment, but the concept should be kept in mind even then. The concept becomes critical when a company (or an individual) has an investment which will significantly change the way income taxes are calculated. Example 1-4 demonstrates the importance of using incremental analysis for a single project.

Example 1-4
The incremental analysis concept applied to a single project.

A project has a net operating income before income taxes of $60,000 in a given year. In the same year the allowable deductions for calculating income taxes are $110,000 resulting in a taxable income of $50,000. Let's evaluate the project for that year without using incremental analysis.

SIMPLIFIED METHOD
(Not always correct)

	NOIBFIT from project	$ 60,000
less	Deductions for FIT	110,000
equals	Taxable Income	$– 50,000

The Cash Tax liability at a 46% rate is $— 23,000

	NOIBFIT	$ 60,000
less	cash taxes	– 23,000
equals	NOIAFIT	$ 83,000
less	investments	0
equals	NCFAFIT	$ 83,000

Now let's look at the same project using an incremental analysis for two cases—a corporation with a large taxable income, and a corporation with a small taxable income.

Example 1-4
(Continued)

CORRECT APPROACH
LARGE Taxable Income

	Corporation with Project	Corporation w/o Project	Difference is Project
NOIBFIT	$1,060,000	$1,000,000	$ 60,000
Deductions	410,000	300,000	110,000
Taxable Income	650,000	700,000	– 50,000
Income Tax	279,750	302,750	– 23,000
NOIBFIT	1,060,000	1,000,000	60,000
Income Tax	279,750	302,750	– 23,000
Investments	0	0	0
NCFAFIT	780,250	697,250	83,000

This is the same answer we calculated by looking at the project alone. So, in this case, the simplified approach works well.

Now let's examine a corporation with a small taxable income.

CORRECT APPROACH
SMALL Taxable Income

	Corporation with Project	Corporation w/o Project	Difference is Project
NOIBFIT	$210,000	$150,000	$ 60,000
Deductions	210,000	100,000	110,000
Taxable Income	0	50,000	– 50,000
Income Tax	0	9,250	– 9,250
NOIBFIT	210,000	150,000	60,000
Income Tax	0	9,250	– 9,250
Investments	0	0	0
NCFAFIT	210,000	140,750	69,250

The project net cash flow AFIT is $13,750 less for a corporation with a small taxable income than for a corporation with a large taxable income! This is due to the graduated tax rates.

Example 1-4
(Continued)

In order to calculate income taxes for Example 1-4, the following table was used:

Tax Rate	Applied to
17%	1st $25,000
20%	next $25,000
30%	next $25,000
40%	next $25,000
46%	the rest

This example should demonstrate the value of doing a true incremental analysis any time a large expenditure is being evaluated on an after tax basis. Incremental analysis *must* be used when an individual is evaluating a tax shelter. The IRS regulations contain many traps for unwary individuals. Do not assume that since you are in the 50% tax bracket a $1,000 write off will reduce your income tax by $500. Limitations on investment interest deductions, items of tax preference, and other quirks of the IRS code may change the picture.

1.7 UNCERTAINTY (a Form of Risk)

In most of this book we will assume that we are dealing with certainty. i.e. our projections are correct. There are many methods of dealing with uncertainty, but they all first require the ability to evaluate projects assuming certainty. This text is concerned with getting the correct values to put into a risk analysis.

1.8 CASH FLOW DIAGRAM FOR AN OIL PROPERTY

The detailed cash flow diagram for a domestic oil property is shown as Figure 1-6. In evaluating an oil property, each of the items is calculated for each year. The diagram is presented here to form the basis for detailed discussions in later chapters and to define the terms that we will be using in this book. All values below the dashed line are net numbers. That is, the numbers are net revenues and costs for the economic interest being evaluated. Revenues and costs assigned to other parties are not relevant for your project cash flow analysis.

Many times, primarily for independent oil companies, an evaluation will be done on a before income tax basis. The diagram remains the same except that there is no income tax and the net cash flow should be called Net Cash Flow BFIT.

Figure 1-6

Oil Property Cash Flow Diagram

Gross Production
  – Shrinkage
= Gross Sales
  * Net Revenue Interest
= Net Sales
  * Price
- -
= Gross Revenue
  – Severance & Advalorem Taxes
  – Windfall Profits Tax
  – Operating Costs
  – Loan Interest Payments
= Net Operating Income BFIT
  – Federal Income Tax
= Net Operating Income AFIT
  – Investments
  – Loan Principal Payments
= Net Cash Flow AFIT

PROBLEMS

1-1 An oil field transfer pump having a useful life of five years is purchased for $1,000. The pump will generate an incremental $1,500 in the first year. However the revenue will decrease $100 per year in each of the next four years. Operating costs will be $500 in the first year but will increase $100 per year in each of the following four years.

a. Calculate financial net income before income taxes for each year using straight line depreciation (no salvage value).
b. Prepare a net cash flow analysis for this investment. Draw the cash flow diagram.

1-2 For the pump in problem 1-1.

a. Calculate the financial net income after federal income taxes. Use cash taxes in the financial model to calculate after tax net income. Cash taxes are based on straight line depreciation with a 4 year life and 40% tax rate. Financial depreciation is based on a 5 year life straight line method.
b. Calculate the NCF for this project after federal income taxes. Draw the cash flow diagram.

1-3 Calculate NCF using the cash flow model for an oil property evaluation for the following case:

Time Period: 1 year
Gross Production: 50,000 bbl.
Shrinkage: 5%
Sales price of oil: $30/bbl.
Net Revenue Interest: 0.80
Your share of state and local taxes: $90,000
Your share of investments during the year: $100,000
Your share of windfall profits tax: $200,000
Your share of operating costs: $25,000
Your share of federal income taxes: $200,000

1-4 a. A pumping unit costing $30,000 is expected to produce 120,000 barrels before the pumping unit is abandoned. During the first year 40,000 barrels are produced. Calculate units-of production depreciation for the first year.

b. During the second year 50,000 barrels are produced. However, because of an overload placed on the unit it is expected that the unit will be capable of lifting only 10,000 barrels effective the end of the second year. Calculate units-of-production depreciation for the second year.

c. During the third year 15,000 barrels are lifted by the pumping unit before the unit is abandoned. Calculate units of production depreciation for the pumping unit for the third year.

1-5 Calculate financial net income AFIT and net cash flow AFIT for *year 3 only*. Your Time 0 investment for a pumping unit is $50,000. Annual gross revenues are $40,000. Direct costs are $5,000 per year. Depreciation for financial net income is calculated assuming a 5 year life, no salvage value, and the straight line method. Cash income taxes are 40% of *gross* revenue. Assume cash taxes for your financial net income model.

1-6 A. The reservoir pressure in the Bogus Field has declined to the point that Well 1-1 will no longer flow naturally. As a result, it has been decided that a downhole electric pump having a service life of six years will be purchased for $12,000. The pump will cause the well to produce the oil production as shown in the following table.

Year	Production (Bbls)
1	1,000
2	950
3	900
4	850
5	800
6	750

Operating costs are estimated to be $7,500 the first year and will increase by $500 each of the following five years. The pump has no salvage value. Calculate gross revenue using a constant oil price of $20/Bbl.

a. Calculate the financial net income BFIT for each year using straight line depreciation. Assume a six year life.

b. Calculate the financial net income BFIT for each year using UOP depreciation.

c. Prepare a net cash flow analysis BFIT for this investment. Draw the cash flow diagram.

B. If the tax rate is 46% for this project

a. Prepare a net cash flow analysis AFIT for this investment. For income tax purposes, use straight line depreciation assuming a five year life. Prepare a net cash flow diagram.

b. Calculate the financial net income AFIT for each year using straight line depreciation. Assume a six year life for financial depreciation purposes. Use the cash taxes from part a for the financial model.

c. Calculate the financial net income AFIT for each year using UOP depreciation. Again use cash taxes from part a.

Table of Contents
Chapter 2
Time Value of Money

2.1 **Cash Flow Diagram** 2-1
2.2 **Symbols** 2-2
2.3 **Simple Interest** 2-2
2.4 **Compound Interest** 2-3
2.4.1 Future Worth of Lump Sum 2-3
2.4.2 Present Worth of a Lump Sum 2-3
2.4.3 Sinking Fund 2-4
2.4.4 Annuity (Loan) 2-5
2.4.5 Annuity Due or Lease 2-5
2.4.6 Modified Sinking Fund 2-6
2.4.7 Rule of 72 2-6
2.4.8 Nominal and Effective Interest Rates 2-7
2.4.9 Partial Period Discounting 2-8
2.4.10 Changing the Form of the Cash Flow Stream 2-9
2.4.11 Discounting Bonds 2-9
2.5 **Rule of 78** 2-11
2.6 **Compound Interest Equations and Cash Flow Patterns** 2-13

Chapter 2

Time Value of Money

Money has a time value. That means that a dollar received today has more value to us than a dollar received far in the future. This can be demonstrated by offering someone a choice between receiving $100 today and receiving $101 one year from today. Most people will opt for the $100 today. If we increase the amount of future money to $115 or $125 or perhaps $200 and guarantee payment, there will be a point at which the future sum of money will become more attractive than the current $100. The amount of future money necessary to sway the person to choose the future sum is dependent upon many things—the inflation rate, current opportunities to invest the $100, and the perceived risk, among others. No matter what the amount of money necessary to tip the scales, the *concept* that money has a time value is established.

In the case above, if the person is indifferent to receiving $125 one year from now or $100 now, then we say that the two sums are *equivalent*. This concept of equivalence is fundamental to the evaluation of all engineering projects. We are often faced with the choice of having a certain sum of money now or receiving various sums of money over time. By determining the equivalence between money received today and money received in the future, we can make an informed decision.

In the above example, making the choice is relatively simple. There are only two sums to compare and the time period is one year. This is usually not the case, however, so we need a mechanism to handle complex choices. The mechanism that works best is "interest". We can define interest as the amount of money which must be added to our current sum to make an equivalent future sum. The amount of interest necessary to create equivalence is dependent upon the period under consideration. We may be indifferent to receiving $100 now, $112.50 six months from now, or $125 one year from now. In that case, the $12.50 or $25.00 is the amount of interest. In order to easily compare all three alternatives, interest is compared as a rate. It may be expressed as 12.5% per six months or 25% per year. The interest rate is calculated by dividing the amount of interest paid per period by the principal amount at the beginning of the period. Often, interest rate is expressed on an annual basis such as 25% "per annum." It will be easier to understand the formulas which will be developed later, however, if interest rate is thought of as a rate per period such as 1% per month or 6% per quarter. The more common ways of expressing interest rate (nominal and effective annual rates) will be discussed later in this chapter.

2.1 Cash Flow Diagram

One of the most helpful techniques for unraveling complicated interest rate problems is the technique of using cash flow diagrams. A cash flow diagram is a graphical representation showing when the cash flows into and out of a project. An example is shown below:

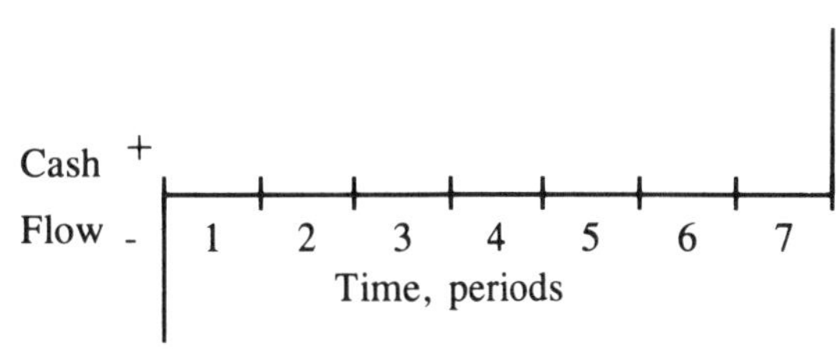

In this example, a sum of money is invested (paid out) at time 0. A different sum of money is received after seven periods. Depending upon the interest rate per period, these sums of money may or may not be equivalent.

Sometimes money is paid out and received at the same time. Although the two sums could be subtracted and the net result plotted, it is often convenient to show the diagram as follows:

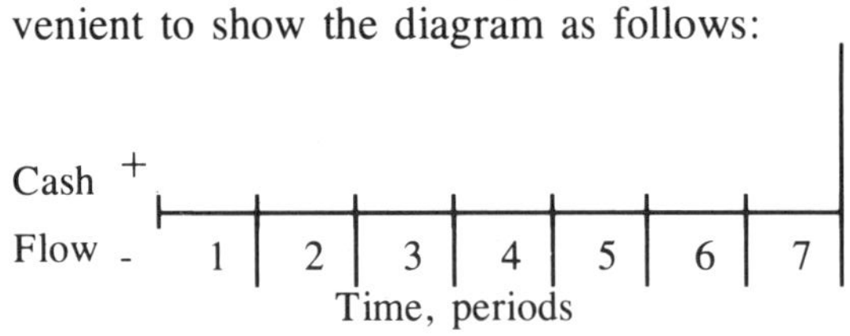

This diagram shows a sinking fund, a concept which will be discussed later.

The importance of these diagrams cannot be overemphasized. By plotting the diagram, the most garbled of alternatives can be understood.

Now that we have introduced the concept of interest, it is time to develop the equations used to make equivalence calculations. The symbols which will be used throughout the rest of this chapter are defined in the following section.

2.2 SYMBOLS

- i = the interest rate *per period*
- n = the total number of interest periods (n = m*t)
- P = a present lump-sum of money
- F = a future lump-sum of money (received **n** periods from now)
- A = an *end*-of-period payment or receipt in a uniform series continuing for **n** periods
- B = a *beginning*-of-period payment or receipt in a uniform series continuing for **n** periods
- m = the number of compounding periods in a year
- t = the number of years **(n=mt)**
- i_n = the nominal annual interest rate ($\mathbf{i_n}$ = **i*m**)
- i_e = the effective yearly interest rate derived from compounding the periodic interest rate **i, m** times

2.3 SIMPLE INTEREST

Simple interest is a concept that is seldom used for periods greater than one year. If an amount **P** is invested at interest rate **i** for **n** periods then the interest will be **P*i*n** and the total amount at the end of **n** periods will be **P*(1 + in).** Many bank loans use this concept for periods less than one year. The formula for the amount **F** obtainable in **n** periods is

$$F = P(1 + in) \qquad (2\text{-}1)$$

Notice that "**n**" does not have to be integer. It can be, and often is, less than 1.

Example 2-1
Simple Interest

$10,000 is borrowed from a bank for one year at 16% per year "simple" interest. If the loan is paid off after 6 months, what is the payoff amount?

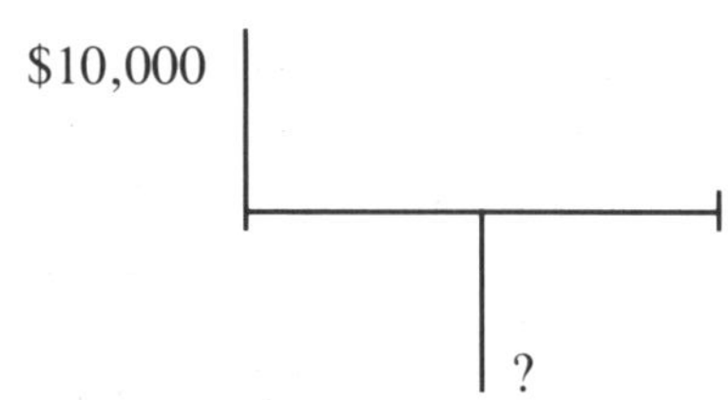

P = $10,000
i = .16/year
n = .5 years
F = P(1 + in) (2-1)
F = $10,000 [1 + (.16*.5)]
F = $10,800

Contrast this to the "Rule-of-78" which appears later in the chapter.

2.4 COMPOUND INTEREST

For loans or bank deposits which exceed one year, the interest is usually "compounded." That is, the interest earned during the first period is added to the original principal in order to form the principal for the second period. The compound interest concept will be used when calculating the equivalence between a sum of money today and future sums of money.

2.4.1 *Future Worth of a Lump Sum*

The future worth of a present sum **P** earning compound interest can be derived by applying the simple interest formula over and over again. Let us add a subscript to the term **F** to indicate that we are talking about the value at the end of the number of periods shown by the subscript. If **P** is invested at **i** (% / **period**) for one period then at the end of that period the amount of money in the account will be:

$$F_1 = P(1+i) \qquad (2\text{-}2)$$

At the end of the second period, interest will be paid on the amount of money which was in the account during the second period, which is $\mathbf{F_1}$ (the original principal plus the first period's interest). Therefore at the end of the second period the amount in the account is:

$$F_2 = F_1(1+i) \qquad (2\text{-}3)$$

At the end of the third period, interest is paid on the amount of money which was in the account during the third period ($\mathbf{F_2}$). The amount in the account at the end of the third period is:

$$F_3 = F_2(1+i) \qquad (2\text{-}4)$$

This process is continued until the money is withdrawn. If the expression for $\mathbf{F_2}$ (eq. 2-3) is substituted into equation 2-4 the result is:

$$F_3 = F_1(1+i)(1+i) \qquad (2\text{-}5)$$

Substituting equation 2-2 for $\mathbf{F_1}$ results in:

$$F_3 = P(1+i)(1+i)(1+i) \qquad (2\text{-}6)$$

or

$$F_3 = P(1+i)^3 \qquad (2\text{-}7)$$

The general equation is:

$$F_n = P(1 + i)^n \qquad (2\text{-}8)$$

The term $\mathbf{(1 + i)^n}$ is called the single payment compound amount factor in many texts and is often tabulated. To emphasize the concepts of time value of money, only the equations will be utilized in this text. No tables of discount factors will be presented.

Example 2-2
Future Value

$1,000 is placed in a bank paying 12% per compounding period. How much money will be in the account after 5 periods?

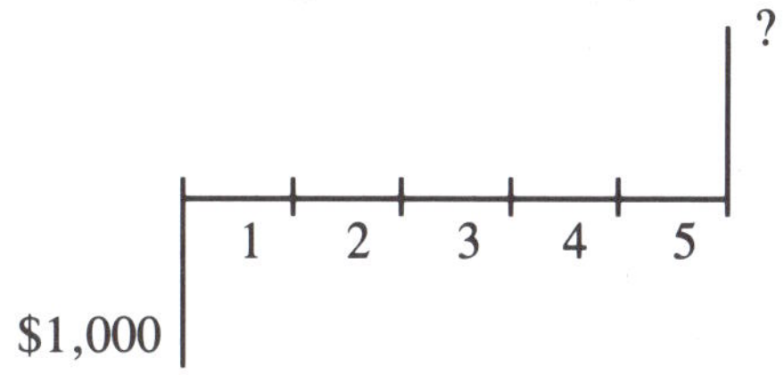

$P = \$1,000$
$i = .12$
$n = 5$
$F = P(1+i)^n \qquad (2\text{-}8)$
$F = \$1,000\ (1.12)^5$
$F = \$1,762$

2.4.2 *Present Worth of Lump Sum*

This is by far the most important equation in this chapter. In fact, by knowing this equation and using the time diagrams mentioned earlier, *any* compound interest problem can be solved.

If an amount **F** is going to be received **n** periods in the future, then its present value **P** can be calculated for a given interest rate **i** using the formula

$$P = \frac{F}{(1 + i)^n} \qquad (2\text{-}9)$$

This is the inverse of the single payment compound amount formula. The justification for this formula lies in the equivalence concept. The sums **P** and **F** are equivalent to each other since **P** could be invested at **i** for **n** periods to become **F**. The term $(1 + i)^{-n}$ is often referred to as the single payment present worth factor or discount factor.

Example 2-3
Present Value

$1,762 is going to be received 5 periods from now. What is the present value of this amount at an interest rate of 12% per period?

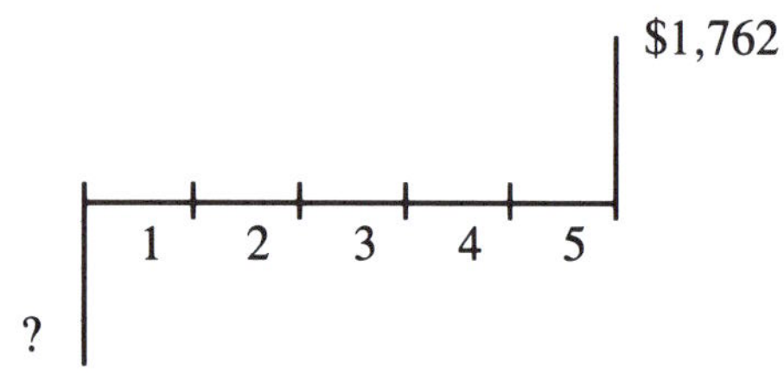

F = $1,762
i = .12
n = 5

$$P = \frac{F}{(1 + i)^n} \qquad (2\text{-}9)$$

$$P = \frac{\$1,762}{(1.12)^5}$$

P = $1,000

2.4.3 *Sinking Fund*

Rather than a lump sum payment as we have been talking about, a sinking fund involves a series of equal payments (**A**) deposited at the end of each compounding period. After **n** periods, the amount **F** is available for withdrawal. This could be likened to a savings plan or retirement fund in which a sum of money is deposited at the end of each period. Since the same amount of money is deposited at the end of each period, this concept is often referred to as a uniform periodic series. The formula relating these terms is:

$$A = F\left[\frac{i}{(1 + i)^n - 1}\right] \qquad (2\text{-}10)$$

The factor $\left[\frac{i}{(1 + i)^n - 1}\right]$ is called the sinking fund factor and will appear often in the rest of the chapter.

Example 2-4
Sinking Fund

One of the American dreams is to become a millionaire. How much money must be deposited at the end of each month for 20 years in an account earning 1% per month compounded monthly to achieve the goal?

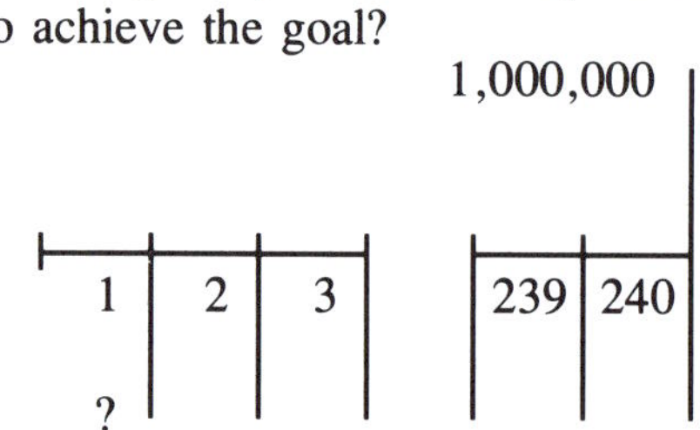

F = $1,000,000
i = .01
n = (20)(12) = 240
A = ?

$$A = F\left[\frac{i}{(1 + i)^n - 1}\right] \qquad (2\text{-}10)$$

$$A = \$1,000,000\left[\frac{.01}{1.01^{240} - 1}\right]$$

A = $1,011 per month

If the account earns 1.5% per month compounded monthly, how much must be deposited?

F = $1,000,000
i = .015
n = 240

$$A = F\left[\frac{i}{(1+i)^n - 1}\right] \qquad (2\text{-}10)$$

$$A = \$1{,}000{,}000\left[\frac{.01}{1.015^{240} - 1}\right]$$

$$A = \$433 \text{ per month}$$

2.4.4 *Annuity (Loan)*

In an annuity, a present amount **P** is exchanged for a uniform series of end-of-period amounts **A** spread over **n** periods. A loan (from the borrower's viewpoint) is just the inverse of this—an amount **P** is received now and payments of **A** are made for **n** periods to pay back the loan. The interest rate charged on the loan is **i** (% per period). The equation relating these factors is:

$$A = P\left[\frac{i}{(1+i)^n - 1} + i\right] \qquad (2\text{-}11)$$

The factor $\left[\frac{i}{(1+i)^n - 1} + i\right]$ is often called the capital recovery factor. Notice it is the sinking fund factor plus the interest rate. The factor is sometimes expressed in the form:

$$\left[\frac{i(1+i)^n}{(1+i)^n - 1}\right]$$

These two forms are equivalent.

Example 2-5
Loan

What are the annual payments on a $10,000 loan with a term of 5 years at an interest rate of 16% per annum compounded annually?

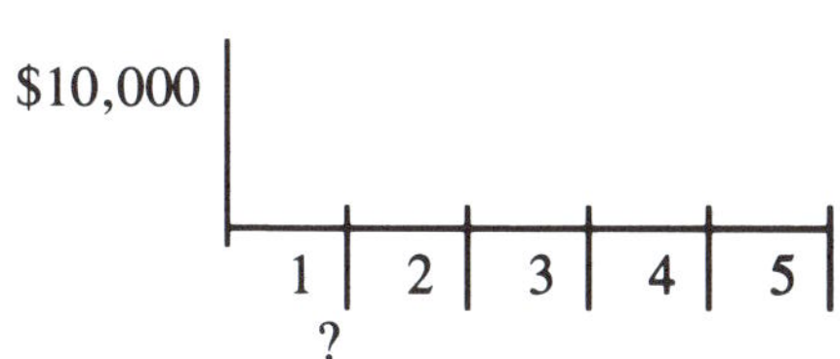

P = $10,000
i = .16
n = 5

$$A = P\left[\frac{i}{(1+i)^n - 1} + i\right] \qquad (2\text{-}11)$$

$$A = \$10{,}000\left[\frac{.16}{1.16^5 - 1} + .16\right]$$

$$A = \$3{,}054$$

2.4.5 *Annuity-Due or Lease*

A lease or annuity-due is the same as a regular annuity, except that the payments are due at the *beginning* of the period. The equation relating **B** and **P** is:

$$B = P\left[\frac{i}{(1+i)^n - 1}(1+i)^{n-1}\right] \qquad (2\text{-}12)$$

Example 2-6
Lease

A student has a 6 month lease on an apartment at $400/month. He would like to pay for the lease in one lump sum. If money is worth 2% per month to the student, how much should he offer to pay the landlord?

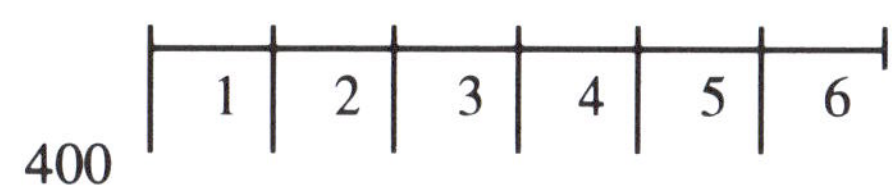

or

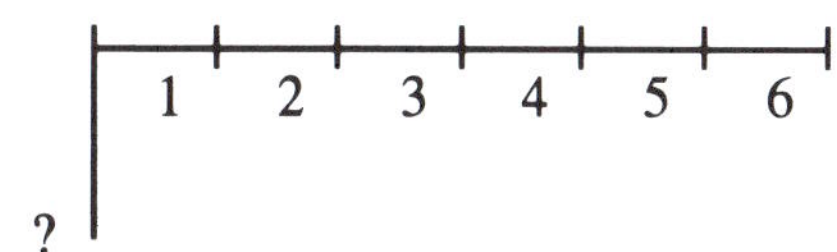

B = $400
i = .02
n = 6

$$P = \frac{B}{\left[\frac{i}{(1+i)^n - 1}(1+i)^{n-1}\right]} \qquad (2\text{-}12)$$

$$P = \frac{400}{\left[\frac{.02}{1.02^6 - 1}(1.02)^5\right]}$$

$$P = \$2,285.38$$

2.4.6 *Modified Sinking Fund*

Often, when saving money, the payments are made at the beginning of the period instead of the end. This is called a modified sinking fund. The relationship between **B** and **F** is:

$$B = F\left[\frac{i}{(1 + i)^n - 1} / (1 + i)\right] \qquad (2\text{-}13)$$

2.4.7 *Rule of 72*

The rule of 72 is an approximation used to determine how long it takes to double your money. If money can be invested at **i**% per period, then it takes **72/i** periods to double or, in equation form,

$$i^*n = 72 \qquad (2\text{-}14)$$

This approximation is surprisingly close throughout a reasonable range of interest rates. Table 2-1 shows the error involved in using the approximation.

Example 2-7
Modified Sinking Fund

How much money must be deposited at the *beginning* of each month to become a millionaire in 20 years if the account pays 1% per month compounded monthly?

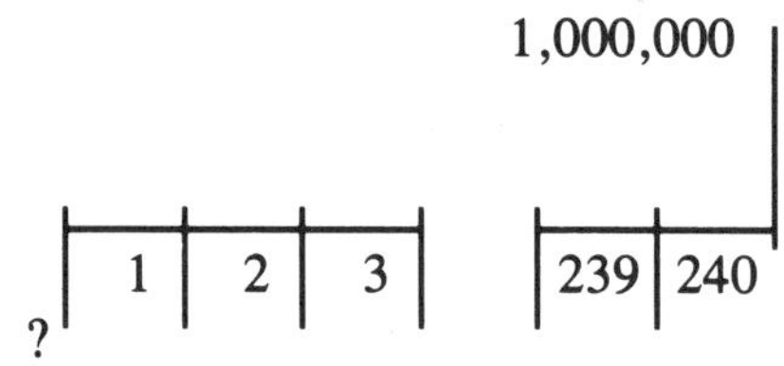

$F = \$1,000,000$
$i = .01$
$n = (20)(12) = 240$
$B = ?$

$$B = F\left[\frac{i}{(1 + i)^n - 1} / (1 + i)\right] \qquad (2\text{-}13)$$

$$B = \$1,000,000\left[\frac{.01}{1.01^{240} - 1} / 1.01\right]$$

$B = \$1,001$ per month

Table 2-1
Error in the Rule of 72

Interest Rate %	n True Time Periods	n^1 Rule of 72 Periods	% error $\frac{n^1 - n}{n} * 100$
1	69.66	72.0	3.4
5	14.21	14.4	1.3
8	9.01	9.0	- .1
10	7.27	7.2	- .9
15	4.96	4.8	-3.2
20	3.80	3.6	-5.6
25	3.11	2.9	-7.3

Example 2-8
Rule of 72

At 18% interest per period, how long does it take to double your money?

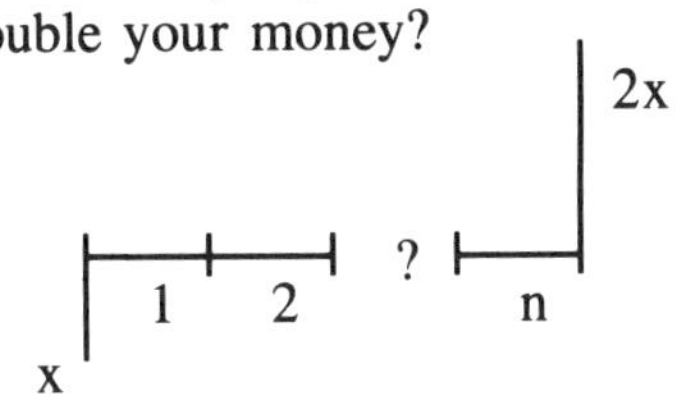

$i = 18$

$n = 72/18 = 4$ periods

2.4.8 *Nominal and Effective Interest Rates*

Interest rates are usually expressed as nominal annual rates instead of the "per-period" rates which we have been using. A nominal annual rate is the product of the period-rate and the number of compounding periods per year.

$$i_n = i * m \qquad (2\text{-}15)$$

The Annual Percentage Rate (APR) required by the Truth in Lending Act is a nominal annual interest rate. For a house or car loan the monthly interest rate is multiplied by 12 to calculate APR.

Effective rate is the yearly interest rate that is equivalent to compounding the period rate (**i**) **m** times per year. The effective interest rate (**i_e**) for any period rate **i** compounded **m** times per year can be determined from the formula:

$$i_e = (1 + i)^m - 1 \qquad (2\text{-}16)$$

or

$$i_e = (1 + \frac{i_n}{m})^m - 1 \qquad (2\text{-}17)$$

Table 2-2 shows the effect of compounding periods other than one year on the effective rate. It can be seen that a compounding frequency greater than monthly has little effect on the effective interest rate. In this book we will normally use one compounding period per year. The use of compounding frequencies other than yearly is exactly equivalent to using the effective interest rate and a period of 1 year in the equations developed in this chapter as long as funds are received once a year—not continuously. It is felt that the continuous flow of funds model does not add to the understanding of oil properties so it will not be developed.

Occasionally, interest is thought of as compounding continuously. In that case the equivalent interest rate can be derived as follows:

$$i = i_n/m \qquad (2\text{-}15)$$

$$i_e = (1 + \frac{i_n}{m})^m - 1 \qquad (2\text{-}18)$$

The limit of $(1 + \frac{i_n}{m})^m$ as **m** approaches infinity is e^{i_n} so

$$i_e = e^{i_n} - 1 \qquad (2\text{-}19)$$

for continuous compounding.

Table 2-2
The effect of compounding frequency

m	i	i_n	i_e	$\frac{i_e - 1}{i} * 100$
	%	%	%	%
1	12.00000	12	12.0000	0.0
2	6.00000	12	12.3600	3.0
4	3.00000	12	12.5509	4.6
12	1.00000	12	12.6825	5.7
365	.03288	12	12.7475	6.2
∞	—	12	12.7497	6.2

2.4.9 *Partial Period Discounting*

The discount formulas presented thus far assume that money changes hands at the end of each period or the beginning of each period. Often this is not a good model of how cash actually flows. For instance, in a producing oil property, the cash flows in and out on a monthly basis. If we wish to truly model the flow of income we would show the cash receipt for January crude oil runs as occurring at the end of February. By the time the true monthly expenditure pattern is superimposed, the picture becomes a mess. Further, we usually don't gain much by going to the trouble of forecasting monthly. For these reasons, an oil property is usually evaluated on a yearly basis. The following diagram shows the monthly cash flow pattern associated with a year's oil production.

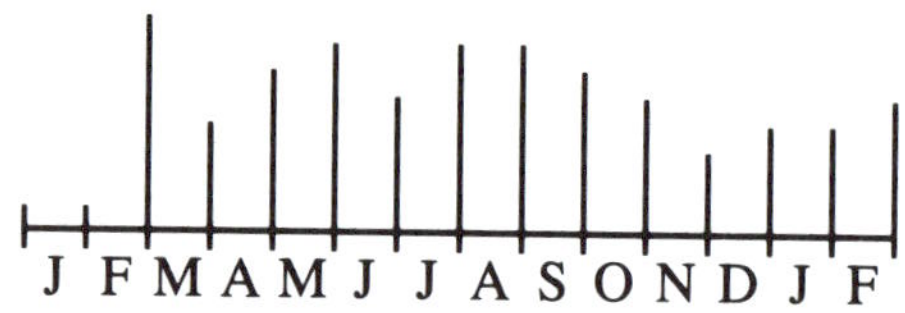

Notice that the cash for January's runs is received at the end of February. When this cash flow pattern is modeled by a yearly lump sum, it appears that a mid-year sum would approximate the monthly cash flows better than an end-year sum. Therefore, most oil properties are evaluated using mid-period discounting rather than end-period discounting. The change in present value between end-year and mid-year discounting is minor when compared to the inaccuracies in the production and price forecast.

Example 2-9
Uneven flow of funds

The following table shows the cash flow from a certain investment. What is the present value of the net cash flow stream at 15% per annum as of January 1, 1983.

$$P = \frac{F}{(1 + i)^n} \tag{2-9}$$

(1) Date	(2) Net Cash Flow $	(3) Years from Time 0* (n)	(4) Present Worth Factor $(1+i)^{-n}$	(5) Present Worth $ (2)*(4)
Jan. 1, 1983	-1,000	.000	1.000	-1,000
July 1, 1983	200	.496	.933	187
Aug. 15, 1983	100	.619	.917	92
Dec. 1, 1983	500	.915	.880	440
July 1, 1984	400	1.499	.811	324
Oct. 1, 1984	800	1.751	.783	626
				669

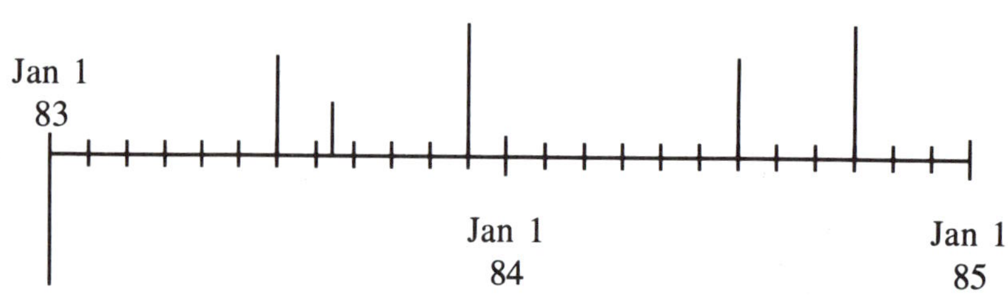

* the timing is calculated based on the number of calendar days from time 0 to the receipt of cash.

The middle of the year is not the only point at which it can be assumed that money is received or paid out. Any assumption can be made if it more closely approximates the true flow of cash. Example 2-9 illustrates a rather strange flow of funds.

Even though this cash flow pattern is quite unusual, we can easily find its present value. Once the time diagram is drawn, it is a simple matter to determine the number of periods from 0 to each cash flow. Since we are compounding annually, the number of periods is equal to the number of years. The present worth factor is then calculated for each cash flow and multiplied by that cash flow to determine its present value. The sum of the present worths of the individual cash flows is the present worth of the cash flow stream.

2.4.10 *Changing the Form of the Cash Flow Stream*

Because of the concept of equivalence, any form of net cash flow—present lump sum, future lump sum, annuity, etc. can be transformed to any other form. For example, the unusual cash flow stream

---

Example 2-10

Transformation of a Net Cash Flow Stream

Having determined that the net present value at 15%/annum of the cash flow stream in Example 2-9 is $669, calculate the equivalent end-of-year uniform payments lasting 2 years.

$P = \$669$

$i = .15$ (must be same as that used in Example 2-9)

$n = 2$

$$A = P\left[\frac{i}{(1 + i)^n - 1} + i\right] \qquad (2\text{-}11)$$

$$A = 669\left[\frac{.15}{(1.15^2 - 1)} + .15\right]$$

$$A = \$412$$

---

of Example 2-9 could be transformed to a uniform annual annuity. The only constraints are that *the periodic interest rate must be the same* and *the size of the periods must be the same.* Example 2-10 illustrates the principle.

This "changing of form" is quite handy when comparing alternatives. If money is worth 15%, then the cash flow stream of Example 2-9 is equivalent to a present lump sum of $669 or two equal year-end payments of $412 each. Examples 2-9 and 2-10 show the power of interest rate calculations. It was relatively easy to determine that three seemingly disparate cash flow streams are equivalent to each other.

2.4.11 *Discounting Bonds*

One practical application of the technique shown above is finding the present value of a bond. Bonds are one means used by corporations and municipalities to raise money. A bond is issued with a $1,000 face value and a term of 10 to 20 years (usually). The bond pays interest semiannually at the "coupon rate" which is a nominal annual rate. For instance, if a 20 year bond has a coupon rate of 12%, then the issuer of the bond will pay the bond holder $60 every six months for 20 years. Since only interest is paid, the original amount of the "loan" ($1,000) is due at the end of the 20 years.

After a bond is issued it may be freely traded. On the secondary market, buyers and sellers agree to a price for the bonds. This price is based on the yield that the investor desires. If the investor desires a yield higher than the coupon rate, then he must pay less than the face value for the bond. Conversely, if he is willing to accept less than the coupon rate, he is prepared to pay more than the face value of the bond. As interest rates fall, bond prices rise. As interest rates rise, bond prices fall.

A bond can be evaluated as an annuity combined with a lump sum payment. The price of a bond on the secondary market is found by calculating the present value of the annuity and adding the present value of the future lump sum. The interest rate used to find the present values is the yield desired by the investor. This interest rate is referred to as the "yield-to-maturity." In the financial papers, bond prices are quoted as a percent of face value. If a newspaper lists a bond price as 101-5/8 then an investor would have to pay $1,016.25 for the bond.

Example 2-11 demonstrates how the price of a bond varies with interest rate.

Example 2-11
Discounting Bonds

A municipality issues a 20 year $1000 face value bond with a coupon rate of 14%. After six months, one interest payment of $70 has been made. If the required yield-to-maturity is 14%, how much is the bond worth?

Annuity

$A = \$70$
$n = 40\text{-}1 = 39$
$i = .14/2 = .07$
$P = ?$

$$A = P\left[\frac{i}{(1+i)^n - 1} + i\right] \qquad (2\text{-}11)$$

$$70 = P\left[\frac{.07}{1.07^{39} - 1} + .07\right]$$

$P = \$928.54$

Lump Sum

$F = \$1{,}000$
$n = 39$
$i = .07$
$P = ?$

$$P = \frac{F}{(1+i)^n} \qquad (2\text{-}9)$$

$$P = \frac{1000}{1.07^{39}}$$

$P = \$71.46$

Value of Bond

P Annuity =	$ 928.54
P Lump Sum =	71.46
Price =	$1,000.00

As long as the required yield is the same as the coupon rate, the price of the bond will always be $1,000.

What is the price of the bond if the required yield is 18%?

Annuity

$A = \$70$
$n = 40\text{-}1 = 39$
$i = .18/2 = .09$
$P = ?$

$P = \$750.79$

Lump Sum

$F = \$1{,}000$
$n = 39$
$i = .09$
$P = ?$

$P = \$34.70$

Value of Bond

P Annuity =	$ 750.79
P Lump Sum =	34.70
Price =	$ 785.49

This represents a 21.5% decrease in the price of the bond for a 28.6% increase in the desired yield to maturity.

What happens to the price of the bond if the required yield falls from 14% to 10%?

Annuity

$A = \$70$
$n = 40\text{-}1 = 39$
$i = .10/2 = .05$
$P = ?$

$P = \$1{,}191.19$

Example 2-11
(continued)

Lump Sum
$F = \$1,000$
$n = 39$
$i = .05$
$P = ?$

$P = \$149.15$

Value of Bond

P Annuity =	$1,191.19
P Lump Sum =	149.15
Price =	$1,340.34

This is a 34% increase in the price of the bond for a 28.6% decrease in the desired yield to maturity.

2.5 RULE OF 78

The rule of 78 is often used to calculate the amount of interest that is due if a loan is paid off before the end of its term. It can also be used to calculate the amount of refund that is due if a prepaid insurance policy is cancelled before the end of the term. The Rule of 78 results in a very high effective interest rate and borrowers should avoid loans with this prepayment provision. Of course, lenders should strive to include the rule of 78 in the loan agreement.

The 78 is derived from the sum of the digits of the months in a year. (1 + 2 + 3 + ... + 11 + 12 = 78). Table 2-3 shows the effect of prepaying a single payment note with a one year term at the end of various months. For each month, the table contains the fraction of interest "earned" by the lender; the prepayment factor which, when multiplied by the loan amount, gives the sum of money the borrower must pay to the lender to completely pay off the loan; and a factor which, when multiplied by the stated interest rate for the loan, will give the actual interest rate.

The rule of 78 can be applied to terms other than 1 year. The sum of the digits **1** through **n** can be found easily through the expression

$$SD(n) = \frac{n * (n + 1)}{2} \tag{2-20}$$

Table 2-3
Prepayment of one year note
using Rule of 78

Payoff occurs at end of Month	Fraction of Interest "Earned"	Prepayment Factor	Multiplier to Convert Stated Loan Int. Rate to True Int. Rate
1	12/78	1+ .154i	1.85
2	(12+11)/78	1+ .295i	1.77
3	(12+11+10)/78	1+ .423i	1.69
4	42/78	1+ .538i	1.62
5	50/78	1+ .641i	1.54
6	51/78	1+ .731i	1.46
7	63/78	1+ .808i	1.38
8	68/78	1+ .872i	1.31
9	72/78	1+ .923i	1.23
10	75/78	1+ .962i	1.15
11	77/78	1+ .987i	1.08
12	78/78	1+1.000i	1.00

The amount of interest or premium ''earned'' in the first month is **n/SD(n)**, where **n** is the number of months in the full term. The cumulative amount earned through the second month is **[n + (n-1)]/SD(n)**. The cumulative amount earned through the third month is **[n + (n-1) + (n-2)]/SD(n)**. This can be generalized to calculate the cumulative amount earned through **j** periods when the full term is **n** periods long.

$$\text{fraction earned} = \frac{j*n - SD(j-1)}{SD(n)} \qquad (2\text{-}21)$$

The rule of 78 can be used to calculate the amount of money to refund on a prepaid insurance policy which has been cancelled before its expiration date. When doing this, the refund is the difference between the amount paid and the amount earned. The fraction to be refunded can be expressed more simply as:

$$\text{fraction to be refunded} = \frac{SD(n-j)}{SD(n)} \qquad (2\text{-}22)$$

Example 2-12
Rule-of-78

$10,000 is borrowed from a bank for one year at 16%. The principal and interest are due in one lump sum at the end of the year. If the borrower elects to pay off the note early, the amount due the lender is figured according to the rule of 78. If the loan is paid off after 6 months, what is the payoff amount?

P = 10,000
i = .16

From the table, the prepayment factor is:
1 + .731i.

F = P(1 + .731i)
F = $10,000 (1 + .731*.16)
F = $11,170

The true interest rate is 1.46 * 16% or 23.4%. Not a bad yield for a 16% loan.

Example 2-13
Refund of prepaid insurance policy
using the Rule of 78

A prepaid insurance policy with a term of 5 years is purchased for $2,000. After 6 months the policy is cancelled. How much is the refund?

$$SD(n) = \frac{n * (n+1)}{2} \qquad (2\text{-}20)$$

$$SD(60) = \frac{60 * 61}{2} = 1830$$

$$SD(5) = \frac{5 * 6}{2} = 15$$

$$\text{fraction earned} = \frac{j*n - SD(j-1)}{SD(n)} \qquad (2\text{-}21)$$

Fraction of premium earned =

$$\frac{6*60-15}{1830} = .1885$$

The amount of the refund is
(1—.1885) * $2,000 = $1,623.

Alternatively

$$SD(60\text{-}6) = \frac{54*55}{2} = 1485$$

Fraction of premium refunded =

$$\frac{SD(60\text{-}6)}{SD(60)} = .8115$$

0.8115 * 2000 = $1,623

If the insurance company had used a linear formula (simple interest), then the fraction earned would have been 6/60 or 10%. The refund would have been $1,800.

2.6 COMPOUND INTEREST EQUATIONS AND CASH FLOW PATTERNS

Future Worth of Lump Sum

$$F = P\,(1 + i)^n$$

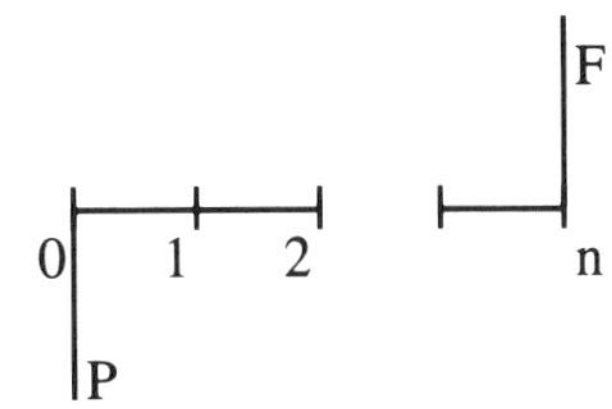

Present Worth of Lump Sum

$$P = \frac{F}{(1 + i)^n}$$

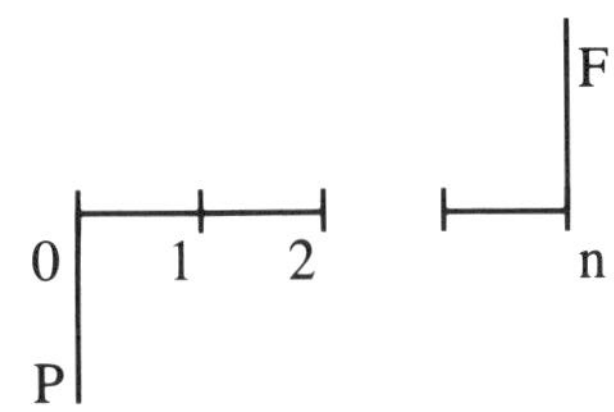

Sinking Fund

$$A = F\left[\frac{i}{(1 + i)^n - 1}\right]$$

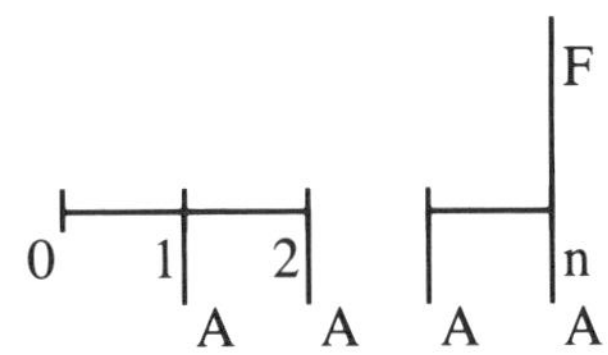

Annuity (Loan)

$$A = P\left[\frac{i}{(1 + i)^n - 1} + i\right]$$

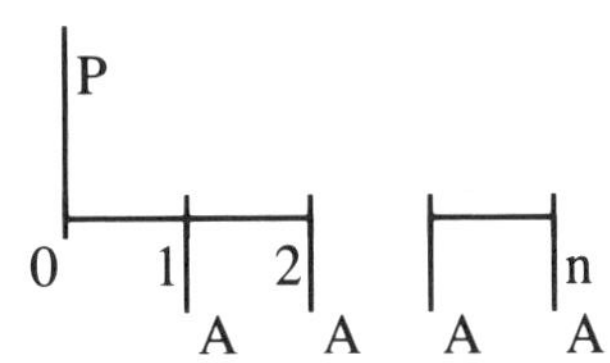

Annuity Due or Lease

$$B = P\left[\frac{i}{(1 + i)^n - 1}\,(1 + i)^{n-1}\right]$$

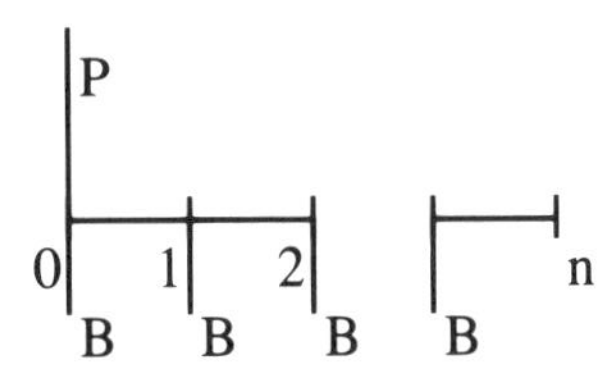

Modified Sinking Fund

$$B = F\left[\frac{i}{(1 + i)^n - 1}\,/\,(1 + i)\right]$$

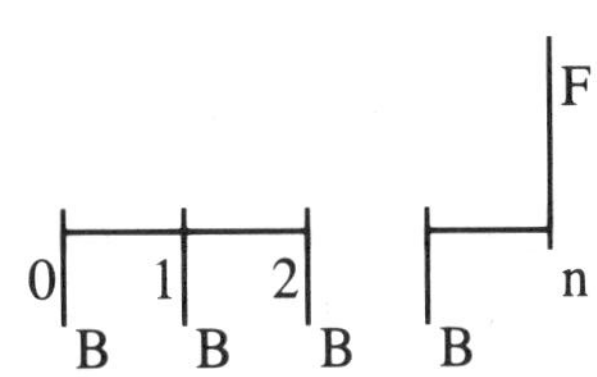

PROBLEMS

2-1 A working interest in a small Kansas oil well is expected to generate the following cash flows at the *end* of each month shown. What would you be willing to pay for the working interest if money has a time value of 1.3333% per month.

Month	Cash Flow
1	−100
2	+500
8	+800
13	−700
16	+1000

2-2 A bank advertizes that it pays an APR of 12%. The same advertisement says that they pay interest at an effective annual yield of 12.616%. How many times per year must the bank be compounding?

2-3 Roughly how many years would it take to turn $1,000 into $16,000 if you can invest the money at 9% per year, compounded annually?

2-4 A "good-deal broker" offers you the following "good-deal." If you give him $200 at the end of each year for the next 20 years he will give you $21,000 at the end of 20 years.

a) If money is worth 15% per year to you, is this a good deal?

b) How much should he give you at the end of 20 years for you to be indifferent?

2-5 You purchase a house and take out a $100,000 loan with a 30 year term at 12% nominal annual interest rate (monthly compounding).

a) What are your monthly payments *to the nearest penny*?

b) If you pay off the loan at the end of 5 years after your 60th payment, how much will you have to pay the bank at that time *to the nearest penny*?

2-6 The installation of a compressor is estimated to generate the following net cash flows. The effective date of the evaluation is May 1, 1984. The projections shown are based on a calendar year.

Year	Net Cash Flow
1 (8 months)	$8,000
2	6,000
3	5,000
4	2,000

a) Using mid-*period* discounting find the present worth of the net cash flows as of May 1, 1984 at 18% compounded annually. Assume all cash flows (in or out) occur at the middle of the *period*.

b) If your company has a time value of money of 18% (nominal annual-compounded annually), should you recommend the compressor installation if it will cost $16,500 (on May 1, 1984)?

c) Using *end*-period discounting what is your recommendation?

2-7 Money is worth 1% per month to you. You're offered the following deals to buy your new BMW 730i.

Deal A:

Pay no money down
Pay $800 per month for 1 year
Pay $1,000 per month for the second year
Pay $1,100 per month for years 3, 4, and 5
You own the car!

Deal B:

Pay X amount down
Pay $1,000 per month for 5 years
You own the car!

What is the amount "X" that you must pay down to make Deal A and Deal B equivalent? Compound monthly.

2-8 a) What is the value of a 20 year $1,000 face value bond with a coupon rate of 8% (pays $40 semi-annually) if the value of money to you is 12%?

Pay $X now
Receive $40 semi-annually for 20 years
Receive $1,000 at end of 20 years

When discounting the $1,000 lump sum payment use semi-annual compounding to agree with the compounding frequency of the uniform series.

b) What is the value of the bond if the value of money to you is 8%?

c) What is the value of the bond if the value of money to you is 16%?

That's the way the bond market works.

2-9 A bank (through its TV ads) is trying to interest you in an IRA (individual retirement account). They say that if you invest $2,000 per year for the next 40 years in their IRA that you will have a nest egg of $600,000 to retire on. What interest rate are they using assuming the money is deposited at the end of the year and that the compounding period is yearly?

2-10 a. What is the value of $1,000 at the end of ten years at 16% nominal annual interest rate?

1) Compounded annually
2) Compounded semiannually
3) Compounded continuously

b. What is the present worth of $1,000 received at the end of the ten years at 16% nominal annual interest rate?

1) Compounded annually
2) Compounded semiannually
3) Compounded continuously

2-11 You want to have $50,000 in the bank after 20 years. If the bank pays 12% compounded *quarterly*, how much must you deposit at the end of each *month* to have this amount of money at the end of 20 years? The bank pays simple interest on money which is deposited during the quarter, but interest is credited to the account and begins to draw interest itself only at the end of a quarter.

2-12 You *deposit* $500 (Time 0) in a bank that pays interest at a 12 ½ % nominal annual rate compounded every four months. A great great aunt is expected to give you $5,000 at the end of 40 months.

You wish to borrow as much as you can at the end of 16 months. You intend to pay off the loan at the end of 40 months with the $5,000 that auntie gives you and the principal and interest from the $500 deposit at time 0. You cannot withdraw the $500 or any interest generated from the bank until the end of 40 months (bank rules won't allow it). The interest rate on the loan is 18% nominal annual rate compounded every four months.

What is the maximum that you can borrow at the end of 16 months?

Table of Contents
Chapter 3
Investment Decision Analysis

3.1 **Types of Decisions** 3-1
3.1.1 Single Project 3-1
3.1.2 Mutually Exclusive Projects 3-1
3.1.3 Portfolio of Projects 3-2
3.2 **Attributes of an Ideal Yardstick** 3-2
3.3 **Yardsticks that Ignore Time Value of Money** 3-2
3.3.1 Urgency 3-2
3.3.2 Payback 3-2
3.3.3 Bail-Out Factor 3-3
3.3.4 Accounting Rate of Return 3-3
3.3.5 Undiscounted Profit to Investment Ratio 3-5
3.4 **Yardsticks That Incorporate The Time Value of Money** 3-5
3.4.1 Net Present Value 3-5
3.4.1.1 Method of Calculation 3-5
3.4.1.2 Concept 3-6
3.4.1.3 Discount Rate Assumption 3-7
3.4.2 Present Value Profile 3-8
3.4.3 Discounted Profit-to-Investment Ratio 3-11
3.4.3.1 Method of Calculation 3-11
3.4.4 Discounted-Cash-Flow Rate-of-Return 3-12
3.4.4.1 Method of Calculation 3-12
3.4.4.2 Concept 3-14
3.4.4.3 Multiple Rates of Return 3-15
3.4.4.4 Future Investments 3-16
3.4.5 Modified DCFROR 3-16
3.4.5.1 Method of Calculation 3-16
3.4.5.2 Concept 3-16
3.4.6 Appreciation of Equity Rate-of-Return 3-21
3.4.6.1 Method of Calculation 3-21
3.4.6.2 Concept 3-22
3.5 **Comparing Alternatives** 3-22
3.5.1 Which Yardstick is Best? 3-25
3.5.1.1 Single Projects 3-25
3.5.1.2 Mutually Exclusive Projects 3-25
3.5.1.3 Capital Budgeting 3-25

Chapter 3

Investment Decision Analysis

Management must make basically three types of investment decisions: (1) A go or no-go decision for a single project, (2) the selection of one project from two or more mutually exclusive projects, and (3) the selection of a portfolio of projects from numerous possibilities. Managers commonly use yardsticks or measures of investment worth to assist them in the decision process. This chapter will present:

(1) a discussion of the three basic types of investment decisions,
(2) the attributes of an ideal yardstick for measuring investment worth, and
(3) a description of some of the investment yardsticks currently in use and their advantages and disadvantages.

3.1 TYPES OF DECISIONS

3.1.1 *Single Project*

The first test that any project must pass is the "go or no-go" criteria established by management. That is, each project must be able to stand alone. Incremental analysis, the key to every investment decision, also applies to single project analysis as discussed in Chapter 1. In this case, the increment being evaluated is the difference between the cash flow of the entire company (or economic entity) including the proposed project, less the cash flow of the entire company without the proposed project. In other words, what incremental cash flows are generated as a result of the incremental investment? Usually, management does not go through this process, even though, as was pointed out in Chapter 1, in some cases such an approach should be used, especially for small corporations and individuals.

In making the go or no-go decision for a single project, management usually calculates several yardsticks to measure profitability. Typically the final decision is not based on a single attribute, but is a combination of factors including good judgement by management. Of all the factors, good judgment by management is probably the single most important factor. This is especially true when you consider the many factors that cannot be quantified.

3.1.2 *Mutually Exclusive Projects*

Mutually exclusive projects are two or more projects where the acceptance of one means the rejection of all of the others. That is, only one of the projects may be undertaken. All the projects being considered must first pass the single project go or no-go decision criteria established by management. In the classical approach, each project that is acceptable on a stand-alone basis is then compared to each of the remaining projects. Thus, again, the incremental concept is applied. The incremental cash flows are computed for each pair of alternatives by subtracting the cash flows of the project having the smaller initial investment from the cash flows of the project with the larger initial investment. If the incremental investment required by the larger project generates incremental cash flows that result in a set of favorable attributes as measured by management's yardsticks, the project requiring the larger investment is selected. Otherwise, the project having the smaller initial investment is selected. In cases where there are more than two projects, the project surviving the comparison is then compared to one of the remaining projects. This process is continued until only one project remains. This long and complicated process can be reduced to a simple comparison of present values as will be shown later.

3.1.3 *Portfolio of Projects*

The selection of a portfolio of projects from numerous possible projects is the decision process during capital budget deliberations. If capital is unlimited all projects meeting the minimum requirements for a single project would be selected. However, this is usually not the case; ordinarily, capital restrictions require that some projects be eliminated. The decision process should select, given the capital constraints, the suite of projects that gives the best combined value for the measure of investment worth being used by management.

The application of various yardsticks to each of these decision types will be discussed later in the chapter.

3.2 ATTRIBUTES OF AN IDEAL YARDSTICK

Since management must constantly make decisions, yardsticks or measures of investment worth are used to help compare and rank competing projects. Investment yardsticks help quantify the investment decision in cases where money, personnel, equipment, and time are limited. Management and financial experts dream of finding an ideal yardstick that will take all the risk out of investment decision analysis. Unfortunately, a single yardstick has not been found; however, it is useful to establish the characteristics of an ideal yardstick. This will permit a better understanding of the limitations and advantages of the various yardsticks that are available. Following is a list of the characteristics of an ideal yardstick:

- Consistent with corporate goals.
- Easy to understand and apply.
- Permits cost-effective decision making.
- Provides a quantitative measure for acceptance or rejection.
- Permits alternatives to be compared and ranked.
- Incorporates the time value of money.

Note that the yardstick must help management quantify the decision process. With these characteristics in mind, the following paragraphs will present some yardsticks that may be used by management.

Two general categories of measures of investment worth will be presented: (1) yardsticks that ignore the time value of money, and (2) yardsticks that incorporate the time value of money. The method of calculation, concept, and the advantages and disadvantages of each measure of investment worth will be discussed.

3.3 YARDSTICKS THAT IGNORE TIME VALUE OF MONEY

Payback or payout, bail-out factor, accounting rate of return, profit to investment ratio, and urgency are five decision models that ignore the time value of money. Despite the fact that these models are fundamentally inferior to the decision models that incorporate the time value of money, it is important to understand them since they are still used by many firms.

3.3.1 *Urgency*

Many times an investment decision is made without methodical analysis because a quick decision is needed either to keep a deal or to get part of the action. Sometimes the results are good, but often the decision is not in the best interest of the firm. The ability to persuade is often a key element in urgent decisions. Since economic contributions are usually secondary, urgent decisions may or may not enhance the future profitability of the firm.

3.3.2 *Payback*

Payback or payout is a measure of the time period required to return the original investment. Although academic writers have generally denounced the use of the payback period as an investment yardstick, surveys show that many managers still use the technique. It is often used as a constraint, rather than as a criterion for optimization, in conjunction with other measures of investment worth. Despite its usage, the payback period method does have several limitations:

- Does not measure profitability
- Ignores all cash flows subsequent to payback point including the salvage value.
- Ignores the time value of money.

Despite the flaws associated with the payback period, payback can be used as a measure of the risk associated with a project. This is especially true for multinational corporations investing in politically unstable countries where the uncertainty about the outcome of an investment must be resolved in a specific amount of time. The payback period, which is the time required for the original investment to be returned, is a measure of when the projects uncertainty is expected to be resolved.

Example 3-1
Using Payout as a Yardstick

Assume that the choice exists between two machines, A and B. Assume that both machines will result in an operating savings of $1000 per year. Machine A costs $5,000 versus $4,000 for machine B. Comparing the payout periods shows that machine A will payout in 5 years (5000/1000) and machine B will payout in 4 years (4000/1000). Using payout as the only criterion, machine B would be purchased. However, if you know that machine A has a useful life of 10 years versus a useful life of only 4 years for machine B, then the decision would be different.

3.3.3 *Bail-Out Factor*

Similar to the payback period but probably a better measure of risk in the sense of avoiding loss is the *bail-out* factor. That is, which project offers the best bail-out conditions if plans go bad? The bail-out payback time is assumed to occur when the cumulative net cash inflow plus the salvage value at a point in time is equal to the original investment. Consider Example 3-2.

3.3.4 *Accounting Rate Of Return*

The accounting rate of return (ARR) yardstick is a financial model that is also known as the book-value rate of return, the approximate rate of return, the average rate of return, the accrual accounting model, and the financial method. The ARR financial model for evaluating investments was widely used prior to 1950 when discounting techniques were first introduced, and despite more sophisticated techniques is still used by many managers today. The ARR method of decision analysis should be understood since it is still used by many managers. Furthermore, decisions made using discounting techniques are sometimes evaluated (post audit) using a financial model. Such a comparison, however, is not valid.

The financial model for making investment decisions assumes the accrual method of accounting and financial income as a measure of profit. The assumption of accrual accounting introduces the following limitations:

- The time value of money is ignored.
- The decision to capitalize or expense an item is somewhat arbitrary.

There is no consistent method to calculate the accounting rate of return. Two equations for the accounting rate of return are:

$$ARR = \frac{NI}{\left(\frac{I - S}{2}\right)} \qquad (3\text{-}1)$$

where: NI = average financial income (after taxes) over the estimated life of the project.
I = initial capital investment
S = salvage value

$$ARR = \frac{NI}{I} \qquad (3\text{-}2)$$

where: I = initial capital investment
NI = average financial income (after taxes) over the estimated life of the project.

In looking at the formulas for the accounting rate of return, it is apparent that the results can vary considerably depending on the assumptions.

Example 3-2
Using bail-out vs payback as a yardstick

Assume that two pumps having the same useful life of 10 years cost $300,000 and $400,000 (Pumps A and B respectively). Pump B costs more because it is a special purpose pump whereas Pump A is a standard pump. In return for the additional cost, Pump B is expected to have net cash inflows of $80,000 per year versus $50,000 (higher maintenance) per year for Pump A. However, because the market demand for Pump B is lower, it has a lower percentage salvage value than Pump A. The salvage value for Pump B is $250,000 at the end of the first year and decreases $25,000 per year thereafter. Pump A has a salvage value of $200,000 at the end of the first year which decreases $20,000 per year thereafter.

Cash Flow Analysis
Pump A
Initial Investment $300,000

Year	Cumulative Net Cash Flow ($M)	Salvage Value ($M)	Cumulative Net Cash Flow + Salvage ($M)
1	50	200	250
2	100	180	280
3	150	160	310
4	200	140	340
5	250	120	370
6	300	100	400

Bail-out period: 2.8 years assuming a salvage value of $160,000
Traditional payout period: 6.0 years (ignores salvage value)

Cash Flow Analysis
Pump B
Initial Investment $400,000

Year	Cumulative Net Cash Flow ($M)	Salvage Value ($M)	Cumulative Net Cash Flow + Salvage ($M)
1	80	150	230
2	160	125	285
3	240	100	340
4	320	75	395
5	400	50	450
6	480	25	505

Bail-out period: 4.4 years assuming salvage value of $50,000
Traditional payout period: 5 years (ignores salvage value)

This example demonstrates that different conclusions can be reached depending on the interpretation of payout.

For example, using the average investment versus the initial investment simply doubles the rate of return. It is such assumptions that make the accounting rate of return model difficult to apply. The question arises, for example, whether or not research and sales promotion costs should be included in the investment base or included as an expense. Working capital items such as cash, accounts receivable, and inventories that increase as a result of the project should be included in the initial investment when calculating the ARR.

Example 3-3
Calculating the Accounting Rate of Return

Using our oil well Example 1-2, calculate the accounting rate of return. The results of Example 1-2 are summarized below:

- Initial Investment = \$60,000
- Financial Net Income over life of project = \$20,430
- Project Life = 6 years

$$ARR = \frac{NI}{\left(\frac{I - S}{2}\right)} \qquad (3\text{-}1)$$

$$ARR = \frac{\left(\frac{20{,}430}{6}\right)}{\left(\frac{60{,}000 - 0}{2}\right)} = 0.114$$

or

$$ARR = \frac{NI}{I} \qquad (3\text{-}2)$$

$$ARR = \frac{\left(\frac{20{,}430}{6}\right)}{60{,}000} = 0.057$$

3.3.5 *Undiscounted Profit to Investment Ratio*

The profit to investment ratio is a quantitative yardstick, but the absolute magnitudes of the profit and investment are not reflected in the ratio. In calculating profit to investment ratios, several variations are available. In one method the profit to investment ratio may be calculated by taking the ultimate after tax net cash flow for the life of the project and dividing this amount by the investment. Another variation is to calculate net operating income (rather than net cash flow) and divide this amount by the investment. The use of net operating income rather than project net cash flow causes the ratio to increase by one. The profit to investment ratio is calculated both ways in Example 3-4.

Another modification to the profit to investment ratio is to calculate the discounted values for both the numerator and the denominator. This has the effect of taking into account the time value of money and the pattern of cash flows. The merits of the discounted profit to investment ratio will be discussed in the next section which presents the techniques that incorporate the time value of money.

3.4 YARDSTICKS THAT INCORPORATE THE TIME VALUE OF MONEY

The current trend is toward yardsticks that incorporate the time value of money. However, prior to 1950 this was not the case. During the 1950's petroleum companies and public utilities began to use yardsticks that incorporated the time value of money, and surveys subsequent to 1960 indicate an increasing use of the discounted yardsticks and a decreasing use of the accounting rate of return and payback methods. The following paragraphs will present some of the yardsticks that incorporate the time value of money.

3.4.1 *Net Present Value*

3.4.1.1 *Method of Calculation*

The net present value of an investment is calculated by discounting the future net cash flows to time zero and summing them. In its simplest form, the equation for the net present value can be expressed as:

$$\sum_{j=0}^{L} \frac{NCF_j}{(1 + i)^j} \qquad (3\text{-}3)$$

where:

L = project life

NCF_j = net cash flow for period j

NPV = net present value discounted at rate i

The above equation assumes that investments subsequent to time 0 are treated as cash out and are included in the annual NCF calculation. It also assumes that all net cash flows occur at the end of the period.

The discussion on the selection of the appropriate rate for discounting the net cash flows will be deferred to the following paragraphs. The meaning of and assumptions inherent in the NPV yardstick will be presented at that time, also.

3.4.1.2 *Concept*

When using the net present value method to measure investment worth, one is applying the concepts developed in Chapter 2 concerning the time value of money and equivalence. What one is asking is this: If money has a time value of

Example 3-4
Profit to Investment Ratio Calculations

Assume that we are still considering investing $60,000 in our oil well (Example 1-1). Calculate the profit to investment ratio using (a) the net cash flow for "profit," and (b) the net operating income for "profit." The cash flow projections are repeated here for your convenience.

Year	Investment	Net Operating Income	Net Cash Flow
0	$ 60,000	—	$ −60,000
1	—	$ 37,100	37,100
2	—	16,800	16,800
3	—	12,200	12,200
4	—	8,640	8,640
5	—	5,440	5,440
6	—	250	250
		$ 80,430	$ 20,430

Net Cash Flow to Investment Ratio

Net cash flow for life of project = $20,430
Investment = $60,000

$$\frac{\text{NCF}}{\text{Investment}} = \frac{20{,}430}{60{,}000} = 0.34$$

Net Operating Income to Investment Ratio

Net Operating Income over Life of Project = $80,430
Investment = $60,000

$$\frac{\text{NOI}}{\text{Investment}} = \frac{80{,}430}{60{,}000} = 1.34$$

Notice that the net operating income to investment ratio can be calculated by adding 1 to the net cash flow to investment ratio. This is true for all cases.

X% to me, would I rather invest a certain amount of money in exchange for an estimated future cash stream, or keep my money? In order to make this decision, all cash flows need to be measured at some common time. Usually this is time 0. By doing this, one can compare the initial investment at time 0 with the present value (discounted at your time value of money) of the future net cash flows. If the present value of the future net cash flows is greater than the time 0 investment, then you would prefer the investment and future cash stream over keeping your money. In the case where your time 0 investment is equal to the present value of the future net cash flow stream, you would be indifferent. When the time 0 investment is greater than the discounted future net cash flow stream, you would prefer to keep your money and look for another opportunity.

3.4.1.3 *Discount Rate Assumption*

In the previous paragraph, the discount rate was assumed to be equivalent to your time value of money. This is analogous to saying that, on average, typical investments should provide this interest rate. In other words, your time value of money could be thought of as your average investment opportunity rate. You probably would not be interested in investing in a project having a negative net present value discounted at your average investment opportunity rate, since by undertaking such an investment, you would forego the opportunity to invest in a project having a greater return (hopefully). Thus it appears, at this point, appropriate to use your average investment opportunity rate to discount the future net cash flows in calculating the NPV. Another approach, used by some, is to discount the future net cash flows at the average cost of funds. This cost of funds is usually referred to as the average cost of capital. Obviously, an investment should return more than the cost of money used to make the investment itself. However, one should really look beyond the cost of capital and take a look at the reinvestment of the net cash that is generated from

Example 3-5
Net Present Value Calculation

Again assume our $60,000 investment in the proposed oil well. The annual net cash flow projections and present value calculations are summarized below:

YEAR	NET CASH FLOW	PRESENT VALUE DISCOUNTED AT 10%
0	$ 60,000	$ 60,000
1	37,100	33,727
2	16,800	13,884
3	12,200	9,166
4	8,640	5,901
5	5,440	3,378
6	250	141
		$ 6,198

Net Present Value for Project

$$NPV = \sum_{j=0}^{L} \frac{NCF_j}{(1+i)^j}$$

NPV = $6,198

a typical investment. To help understand the reinvestment concept and flow of funds, please refer to Figure 3-1.

Notice in Figure 3-1 that there are two sources of funds, an external source and an internal source. However, once money from either source is deposited in the company treasury, it loses its identity. That is, one cannot distinguish a dollar generated by a project (internal source) from a dollar generated through a new debt issue (external source). Thus, a pool of funds has been created. A desirable goal is to make investments that will cause the treasury to grow. In order to do this, one needs to consider what uses are made of the internally generated funds as well as the external source of funds. For example, as internal funds are generated, what investment opportunities exist? By pursuing projects that have a positive net present value discounted at your average investment opportunity rate, the treasury will grow at a faster rate than the assumed investment opportunity rate (assuming that the projects perform as expected, of course). However, if the cost of capital is used to discount the future net cash flows, and projects are selected that have a positive net present value when discounted at your average cost of capital, the treasury *may not* grow as fast as the average investment opportunity rate.

Thus, using this concept of funds and the uses that are made of these funds, one should use the company average investment opportunity rate to calculate the NPV for a project. When this is done, the technique explicitly accounts for the reinvestment of the project's net cash flows when comparing alternatives. The objective in this case is treasury growth. To ensure that the treasury will grow at the fastest possible rate, only projects having a positive net present value when discounted at the company's average investment opportunity rate should be selected.

Example 3-6 demonstrates how selecting projects having the largest NPV when discounted at the average reinvestment opportunity rate will cause the treasury to grow at the fastest possible rate.

3.4.2 *Present Value Profile*

A technique that is valuable in comparing mutually exclusive projects is the *present value profile*. The present value profile is generated by calculating the NPV for each project at several discount rates. The plot of NPV versus discount rate is then made. When several projects are plotted on the same graph, the present value profile plot enables the decision maker to quickly determine the best project among several mutually ex-

FIGURE 3-1

Sources And Uses Of Funds

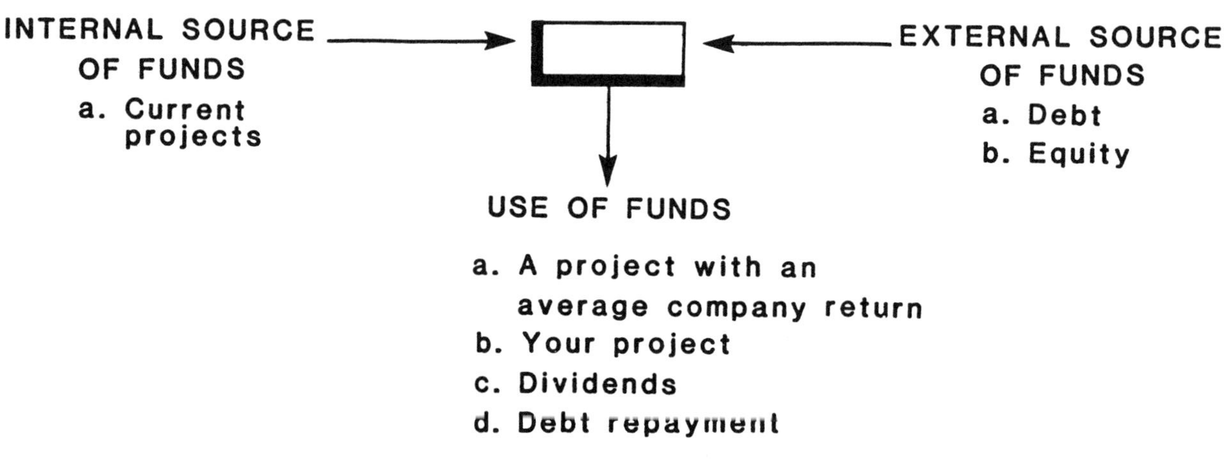

Example 3-6
Discount Rate and the Reinvestment Assumption

Assume that your company treasury has $9,000 available for investments. You have the option to invest in "average" projects that have a 12% return, project A, project B, project C, or any combination of these projects. Assume that average projects are always available in any size that is required. Consider the following cases.

CASE A: Treasury = $9,000 .

Option: Invest in average projects. Average projects have a NPV of 0 when i = 0.12 (DCFROR = 12%). Your reinvestment opportunity rate is 12%.

How much would you have in the treasury at the end of 4 years?

$$F = P\,(1 + i)^n$$
$$F = (9000)(1.12)^4 = \$14{,}161.67$$

CASE B: Treasury = $9,000 .

Option: Invest in average projects and project A.

Project	0	1	2	3	4
A	−1000		1322.50		

How much would be in treasury at end of 4 years?

Project A: $(1322.50)(1.12)^2$	=	$ 1,658.94
Average Projects: $(8000)(1.12)^4$	=	12,588.15
		$14,247.09

CASE C: Treasury = $9,000 .

Option: Invest in average projects and project B.

Project	0	1	2	3	4
B	-3000	3600			

Amount in treasury at the end of four years:

Project B: $(3600)(1.12)^3$	=	$ 5057.74
Average Project: $(6000)(1.12)^4$	=	9,441.12
		$14,498.86

Example3-6
(Continued)

CASE D: Treasury = $9,000

Option: Invest in average projects and project C.

Project	0	1	2	3	4
C	−5000		6050		

Amount in treasury at end of 4 years

Project C: $(6050)(1.12)^2 = \$\ 7589.12$
Average Project: $(4000)(1.12)^4 = 6{,}294.08$
$\$13{,}883.20$

CASE E: Treasury = $9,000 .

Option: Invest in projects A, B, and C.

Project	0	1	2	3	4
A	−1000		1322.50		
B	−3000	3600			
C	−5000		6050.00		

Amount in treasury at end of 4 years.

Project

A $(1322.50)(1.12)^2 = \$\ 1{,}658.94$
B $(3600)(1.12)^3 = 5{,}057.74$
C $(6050)(1.12)^2 = 7{,}589.12$
$\$14{,}305.80$

CASE F: Treasury = $9,000 .

Option: Now, let's invest in project A, B, and an ''average'' project.

Project

A $(1322.50)(1.12)^2 = \$\ 1{,}658.94$
B $(3600)(1.12)^3 = 5{,}057.74$
Average $(5000)(1.12)^4 = 7{,}867.60$
$\$14{,}584.28$

Example 3-6
(Continued)

Finally, let's calculate the NPV @ 12% for project A, B, C and compare the various options available for the $9,000 in the treasury.

$$NPV_A = -1000 + \frac{1322.50}{(1.12)^2} = \$\ 54.29$$

$$NPV_B = -3000 + \frac{3600}{(1.12)^1} = \$214.29$$

$$NPV_C = -5000 + \frac{6050}{(1.12)^2} = \$-176.98$$

For an average project the NPV @ 12% is 0.

COMPARE ALTERNATIVES

Option	Amount in treasury at end of 4 years	NPV @ 12%
Average Projects	$14,161.67	0.00
A + Avg. Project	14,247.09	54.29
B + Avg. Project	14,498.86	214.29
C + Avg. Project	13,883.20	−176.98
A + B + C	14,305.80	91.60
A + B + Avg. Project	14,584.28	268.50

Given that only project A, B, C and ''average'' projects are available and that our goal is treasury growth, we would prefer projects A, B, and a $5,000 average project. This combination results in the largest possible sum in the treasury at the end of 4 years. Note also, that the NPV @ 12% for this option is also the largest. Thus, if one selects projects that maximize the NPV when discounted at the average reinvestment opportunity rate, you will have selected the suite of projects that will cause the sum in the treasury to be maximized (given the constraint of treasury size and project selection). This of course assumes that future cash flows can be reinvested at the average reinvestment opportunity rate.

clusive projects at any assumed investment opportunity rate.

When comparing projects using the PV profile, certain attributes of the cash flow stream can be ascertained. To demonstrate, lets make a present value profile (PVP) plot for two projects in Example 3-7.

3.4.3 *Discounted Profit to Investment Ratio*

The discounted profit to investment ratio is a measure of investment efficiency. Profit can be either defined as Net Operating Income or Net Cash Flow. When the ''profit'' is divided by the amount of the investment, management has a yardstick that measures the amount of present value profit per dollar of investment. This measure becomes valuable in cases where capital is limited.

3.4.3.1 *Method of Calculation*

The discounted profit to investment ratio can be calculated as follows:

$$\frac{P}{I} = \frac{\sum_{j=0}^{L} \frac{NCF_j}{(1 + i_{op})^j}}{\sum_{j=0}^{L} \frac{INV_j}{(1 + i_{op})^j}} \quad (3\text{-}4)$$

or

$$\frac{P}{I} = \frac{\sum_{j=1}^{L} \frac{NOI_j}{(1 + i_{op})^j}}{\sum_{j=0}^{L} \frac{INV_j}{(1 + i_{op})^j}} \quad (3\text{-}5)$$

Where:

NOI_j = net operating income period j

NCF_j = net cash flow period j

INV_j = investment period j

i_{op} = average investment opportunity rate

L = life of the project

The difference between the value calculated by equation (3-5) and the value calculated by equation (3-4) is 1. This is the same difference as discussed in the section on undiscounted profit to investment.

3.4.4 *Discounted Cash Flow Rate Of Return*

The discounted cash flow rate of return (DCFROR) is a widely used yardstick and is the primary measure of investment worth for many firms. The DCFROR yardstick is also called discounted rate of return, internal yield, internal rate of return, effective yield to maturity, return on investment, profitability index, and earning power. DCFROR is widely used since the yardstick is relatively easy to use; and, unlike the NPV method, does not require that a discount rate be established prior to making the calculation. This apparent advantage however, is the reason why the DCFROR has certain limitations when compared to the NPV method. The limitations and assumptions associated with DCFROR will be discussed in the paragraphs that follow.

3.4.4.1 *Method of Calculation*

The DCF rate of return on investment is calculated by discounting the cash flows at various discount rates until the sum of the discounted net cash flows equals zero. The calculation of the DCFROR is a trial and error process. In equation form, the DCFROR can be expressed as:

$$\sum_{j=0}^{L} \frac{NCF_j}{(1 + i)^j} = 0 \quad (3\text{-}6)$$

where:

NCF_j = net cash flow for period j

L = life of project

i = DCFROR

Since net cash flow equals net operating income minus investment for any year an alternative form of the DCFROR expression can be derived by substituting

$$NCF_j = NOI_j - INV_j$$

in equation 3-6. In that case the equation becomes

$$\sum_{j=0}^{L} \frac{INV_j}{(1 + i)^j} = \sum_{j=1}^{L} \frac{NOI_j}{(1 + i)^j} \quad (3\text{-}7)$$

where:

INV_j = investment during period j

NOI_j = net operating income during period j

L = life of project

i = DCFROR

or

$$\sum_{j=0}^{L} \frac{NCF_j}{(1 + i)^j} = I_0 \quad (3\text{-}8)$$

where:

I_0 = the time zero investment

Example 3-7
Mutually Exclusive Projects — Present Value Profile

Prepare a present value profile for the following two projects.

Year	Net Cash Flow A	Net Cash Flow B
0	$-1000	$-1000
1	1200	0
2	0	0
3	0	0
4	0	1873

	NPV @ 0%	NPV @ 10%	NPV @ 15%	NPV @ 18%
A	$200	$ 91	$ 43	$ 17
B	873	279	71	-34

Using the PVP plot the following observations can be made:

Project A: 1. Project A recovers cash rapidly. This can be determined from the PVP since the NPV is not greatly affected by the assumed discount rate. That is, even at high discount rates, the NPV does not decrease appreciably.

2. If the investment opportunity rate is greater than 16% and less than 20%, you would prefer project A.

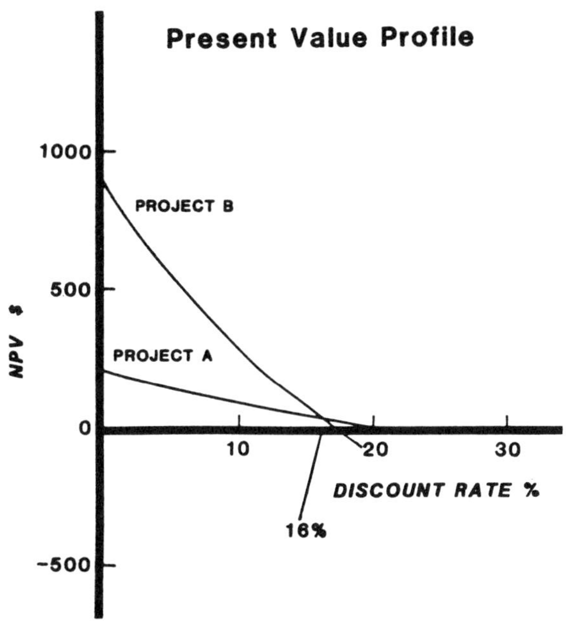

Project B: 1. Project B recovers cash relatively slowly. At low discount rates the NPV is high. However, as the discount rate increases, the NPV decreases rapidly indicating that the present value of cash received in future periods is very sensitive to the assumed discount rate.

2. If the investment opportunity rate is less than 16%, then you would prefer project B.

Also note that if your cost of capital is 12% *and* your average reinvestment opportunity rate is 17%, you would select project A to maximize treasury growth. Thus, if you select projects based on the largest net present value discounted at the cost of capital you would *not* cause the treasury to grow at the fastest possible rate in *this* case.

Each of the above equations can be used to calculate the DCFROR. Example 3-8 demonstrates the DCFROR calculation for our oil well investment.

3.4.4.2 *Concept*

The DCFROR has been described in the literature in many ways but what does it really mean? Supposed assumptions about the reinvestment of cash flows and the idea of an internal rate of return seem to cloud the issue. What then does the calculated rate of return mean? Think of the annuity concept presented in Chapter 2. An annuity does *not* have to be a series of *equal* payments or receipts. In making an investment, you are simply buying an annuity or series of future cash flow streams. Think of making a bank deposit, equivalent to the amount of your investment, in a bank paying interest at a rate equal to the calculated DCFROR. In return, the bank will make a series of payments to you equal to the project's annual net cash flow streams. At the end of the project, your balance in the bank will be zero. Notice, that in the analysis, no mention was made concerning what you do with the money that you receive from the bank. Accordingly, there is *no* assumption about the reinvestment of future cash flows in the DCFROR technique. In the literature, many authors argue, that for the technique to be valid, you must reinvest all the annual net cash flow at the DCFROR or the internal rate of return as it is referred to in the financial literature.

This, however, is not necessarily true. If the annuity concept is accepted, you still earn the calculated DCFROR on your original investment regardless of what you do with the cash flow streams as you receive them. However, when looking at the appreciation of the company treasury and the average investment opportunity rate as we did with the NPV concept, the assumed discount rate becomes, for purposes of comparing alternatives, important. When projects are compared or ranked by using the DCFROR, the ranking is not necessarily equivalent to the ranking one would get using NPV's. Discounting the net cash flows at the average investment opportunity rate takes into account a realistic estimate of the use that can be made of cash as it is received. Thus, if treasury growth is a goal, then accepting projects based solely on DCFROR can lead to decisions that may not be optimal.

Those authors who argue that the DCFROR requires the reinvestment of the net cash flows at the internal rate of return are using treasury growth

Example 3-8
DCFROR Calculation

Using our proposed oil well investment from Example 1-1, calculate the DCFROR. The net cash flow analysis is as follows:

Year	NCF	NPV @ 15.74%		
0	-$60,000	$-60,000/(1.1574)^0$	=	-60,000
1	37,100	$37,100/(1.1574)^1$	=	32,055
2	16,800	$16,800/(1.1574)^2$	=	12,541
3	12,200	$12,200/(1.1574)^3$	=	7,869
4	8,640	$8,640/(1.1574)^4$	=	4,815
5	5,440	$5,440/(1.1574)^5$	=	2,619
6	250	$250/(1.1574)^6$	=	104
		Σ NPV @ 15.74%	$	3

Accordingly, if the discount rate is 15.74%, the net present value is approximately equal to zero and, by definition, this is the DCFROR for this project.

as their corporate goal. Treasury growth is a sinking fund concept — DCFROR is an annuity concept. The problem lies in comparing different concepts. Saying that DCFROR doesn't properly handle reinvestment is like saying that an orange makes a very puny and discolored watermelon.

Attempting to use DCFROR to maximize the rate of treasury growth is a misapplication of the DCFROR technique because, from a treasury growth concept, the interest rate at which the cash flows from the project will be reinvested affects the rate at which the treasury will grow.

3.4.4.3 *Multiple Rates of Return*

Multiple rates of return occur when there are two or more discount rates for which the present value of cash outflows equals the present value of cash inflows. This occurs when the time pattern of net cash flows is such that the sign changes more than once. For example, compare the following cash flow diagrams:

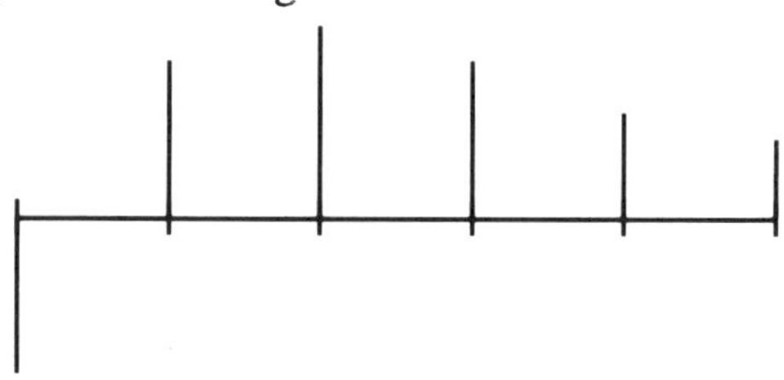

Figure 3-2a

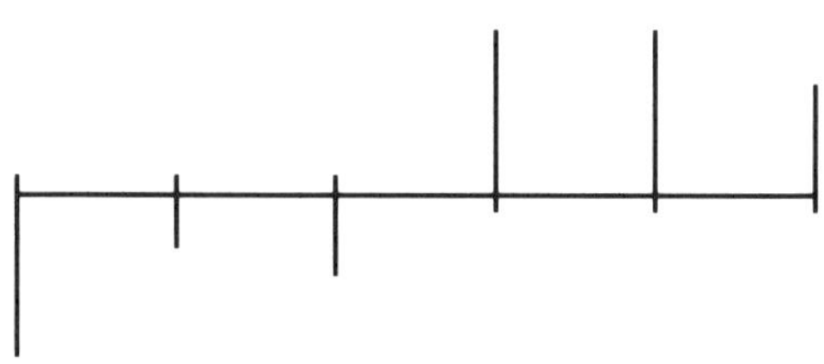

Figure 3-2b

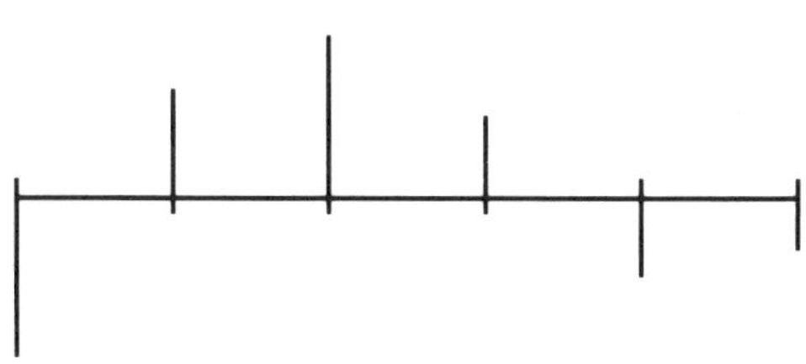

Figure 3-2c

Figure 3-2a is an example of a typical project where an initial investment at time zero resulted in positive net cash flows during the remaining life of the project. Thus, for a project having the estimated net cash flows as shown in Figure 3-2a, the sign changes only once; from a negative sign at time 0 to a positive sign in Year 1. Only one DCFROR will be calculated in this case.

Figure 3-2b is an example of a present value profile for a project requiring several years of lead time before positive net cash flows are generated. An example might be the purchase of an offshore lease (time 0) followed by exploration costs that are needed to determine the viability of the prospect, followed again by the investment for development costs such as platform installation and development wells. Positive net cash flows are finally generated by Year 3. In this example, the sign only changes once (from a negative sign in Year 2 to a positive sign in Year 3). Again, in this situation only one rate of return can be calculated.

Figure 3-2c is an example where the signs do change more than once; from a negative sign at time 0 to a positive sign in Year 1, and from a positive sign in Year 3 to a negative sign in Year 4. In this case, two rates of return are possible. Cash flow patterns where multiple rates of return can occur are found in projects that require large deferred investments such as a waterflood, some incremental analyses of alternative investments, and acceleration projects such as an infill drilling program. An acceleration project simply brings the current NCF stream closer to time 0. This is achieved through an investment. Thus for an acceleration project to be viable, it is important that a good assessment be made of the average investment opportunity rate. In other words, it is important to know at what interest rate the cash from the project can be invested, since the net operating income over the life of an acceleration project is the same as the project before any acceleration of net operating income.

The real question, at this point, is what should one do in the case where multiple rates of return are calculated. One approach is to again use the NPV profile plot. By selecting the project with the largest positive NPV when discounted at the average investment opportunity rate, the problem of multiple rates of return can be resolved. This can best be demonstrated by Example 3-9.

Example 3-9 presents the results of the net present value calculations for various discount rates. These values have been plotted on the present value profile. Notice that the incremental impact of the proposed infill drilling could have been calculated either by subtracting the net present value without the proposed infill drilling from the net present value with the proposed infill drilling, or first subtracting the NCF's and calculating the NPV's for this increment directly.

3.4.4.4 *Future Investments*

Treatment of future investments is another issue that needs to be understood when using the DCFROR technique. Let's consider the case where an initial investment at time 0 is followed by another investment at the end of year 5. In the DCFROR technique, all cash flows are discounted at the DCFROR to time 0. Discounting the year 5 investment to year 0 at the DCFROR implies that the discounted amount could be put in a bank at the DCFROR and a sum equal to our future investment could be withdrawn at the end of year 5. This, however, may not be realistic. A more realistic approach may be to discount the future investments at the average investment opportunity rate and then proceed as before in calculating the DCFROR. This approach makes a realistic equivalence calculation where future investments are involved. The technique will be presented later as the Modified DCFROR.

3.4.5 *Modified DCFROR*

The problem of discounting future investments at the DCFROR rather than at the average investment opportunity rate was introduced earlier. A solution to this problem is presented here as the modified DCFROR.

3.4.5.1 *Method of Calculation*

The modified DCFROR is calculated by discounting all investments to time 0 at the average company investment opportunity rate. After that the method of calculation is the same as the conventional DCFROR technique. The solution, still trial and error, is the interest rate which makes the present value of the net operating income equal to the present value of the investments discounted at the average investment opportunity rate. In equation form, the modified DCFROR can be expressed as:

$$\sum_{j=0}^{L} \frac{INV_j}{(1 + i_{op})^j} = \sum_{j=1}^{L} \frac{NOI_j}{(1 + i)^j}$$

where:

INV_j = investment during period j

NOI_j = net operating income during period j

i_{op} = average investment opportunity rate

L = life of project

i = modified DCFROR

3.4.5.2 *Concept*

The manner in which the traditional DCFROR technique handles future investments should be challenged. The question that must be answered is: Should future investments be discounted at the DCFROR? Discounting at the DCFROR assumes that the equivalent time 0 investment could be deposited in a bank that pays the DCFROR. At the end of the year when the future investment is to be made, the amount of the future investment could be withdrawn from the bank. This assumption, however, is probably not realistic. A more realistic assumption is to discount all the future investments at the average investment opportunity rate. In this case a realistic equivalence for time 0 investments and future investments is calculated.

The modified DCFROR technique results in DCFROR's that are lower than the traditional DCFROR method. A comparison of the two methods is presented in Example 3-11.

Example 3-9
Multiple Rates of Return — Acceleration Projects

You wish to evaluate a proposed infill drilling program. The estimated cash flow without infill drilling and with infill drilling is presented along with the incremental analysis of the proposed program.

Estimated Cash Flow, M$

Year	With Proposed Infill Drilling Curve "A"		Without Proposed Infill Drilling Curve "B"		Proposed Infill Drilling Curve "C"	
	Invest	NOI	Invest	NOI	Invest	NOI
0	400	—	0	—	400	—
1		1500		600		900
2		1000		500		500
3		200		500		-300
4				400		-400
5				400		-400
6				300		-300
	400	2700	0	2700	400	0

Notice that, on an undiscounted basis, the proposed infill program does not generate any additional net operating income. This project would be considered a "pure" acceleration project. The NOI is simply accelerated by the $400M investment. Now, calculate the NPV at various discount rates so that the Present Value Profile can be drawn.

M$

Discount Rate %	NPV Curve A	NPV Curve B	NPV Curve C
0	2300	2700	−400
10	1940	2025	− 85
20	1660	1591	69
30	1437	1295	142
100	625	530	95
150	373	368	5
210	219	281	− 62
250	115	228	−113
300	41	191	−150
350	−15	165	−180

As shown in the following present value profile, acceptance of the proposed investment to accelerate the NOI's is dependent on the assumed investment opportunity rate. If your average investment rate is between 15% and 150%, you would accept the project. Otherwise you would reject the proposed investment.

EXAMPLE 3-9

Present Value Profile
Infill Drilling Proposal

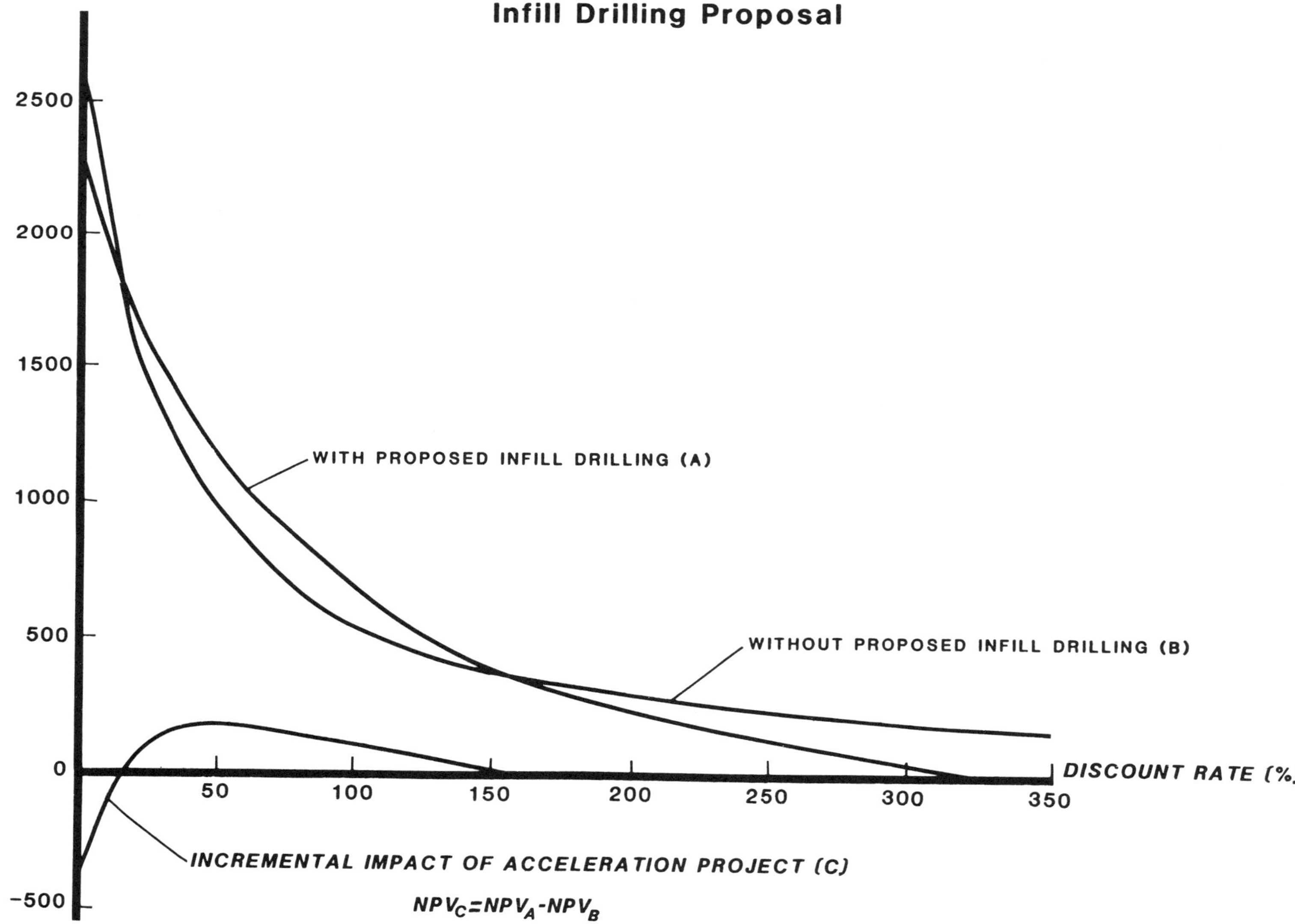

Example 3-10
DCFROR Yardstick — Mutually Exclusive Projects

Compare the following two mutually exclusive projects.

	Net Cash Flow, M$				
	0	1	2	3	4
A	-9000	4000	4000	3000	2000
B	-4500	2000	2000	2000	1000
Incremental	-4500	2000	2000	1000	1000

In this case, the question that must be answered is: Does the incremental investment of $4,500 provide a return equal to or greater than the hurdle rate (possibly your average reinvestment opportunity rate) established by management? In our case, project A and B have DCFROR's of 18.8% and 22.5% respectively. If 10% is the minimum acceptable DCFROR (hurdle rate), both projects are acceptable on a stand alone basis. The incremental investment of $4500 in this case has a DCFROR of 14.8%. Since this return is greater than 10%, the larger investment (Project A) is justified despite the fact that project B has a larger DCFROR.

The present value profile can also be used to assist in comparing two mutually exclusive projects. The discount rate where the two projects have the same net present value is the DCFROR for the incremental cash flows. At discount rates less than the incremental DCFROR, you would pick project A. At discount rates greater than the incremental DCFROR, you prefer project B. These statements assume that each project already meets the minimum hurdle rate established by management.

Present Value Profile

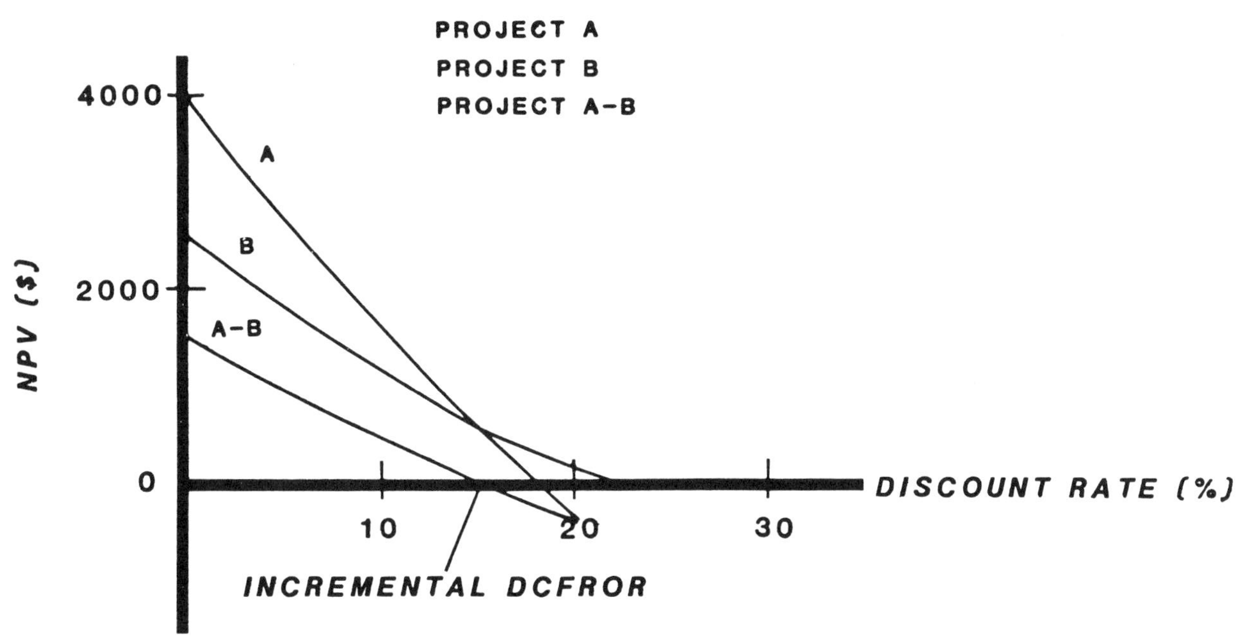

Example 3-11
Handling Future Investments
Modified DCFROR versus DCFROR

Assume that a project has the following cash flow projection:

Year	Investment	Net Operating Income
0	$1000	—
1	0	$500
2	0	300
3	500	400
4	0	1000
5	0	800
	$1500	$3000

Using the conventional DCFROR technique, the DCFROR is 33.7%. The modified DCFROR is 29.3% assuming i_{op} is 15%. Sample calculations are shown below.

Conventional DCFROR:

$$\sum_{j=0}^{L} \frac{NCF_j}{(1+i)^j} = 0$$

$$\frac{-1000}{(1+i)^0} + \frac{500}{(1+i)^1} + \frac{300}{(1+i)^2} + \frac{-100}{(1+i)^3} + \frac{1000}{(1+i)^4} + \frac{800}{(1+i)^5} = 0$$

For our problem, i = 33.7%

Modified DCFROR:

$$\sum_{j=0}^{L} \frac{NOI_j}{(1+i)^j} = \sum_{j=0}^{L} \frac{INV_j}{(1+i_{op})^j}$$

$$\frac{500}{(1+i)^1} + \frac{300}{(1+i)^2} + \frac{400}{(1+i)^3} + \frac{1000}{(1+i)^4} + \frac{800}{(1+i)^5} = \frac{1000}{(1+i_{op})^0} + \frac{500}{(1+i_{op})^3}$$

i_{op} = 15% (average investment opportunity rate)
i = 29.3% (modified DCFROR)

3.4.6 *Appreciation Of Equity Rate Of Return*

The appreciation of equity rate of return (AOEROR) incorporates the concept that the firm's objective is treasury growth. It was originally proposed in the petroleum literature by Phillips (1965). Variations on the theme have been proposed by Capen, Clapp, and Phelps (1973) *(Growth Rate of Return)*, and Roebuck (1984). Here we will present Phillips' version which treats the investment stream separately from the net operating income stream.

3.4.6.1 *Method of Calculation*

The appreciation of equity rate of return is calculated by assuming that all future net operating incomes generated by the investment are invested at the company's average investment opportunity rate. The reinvested net operating incomes are then permitted to grow at an assumed compounding frequency until either a common point in time (for all projects being compared) is reached or the end of the project life is reached. The end-of-project-life approach will be presented later. For now, let's assume all projects will be compared using a common terminal point of "t" years. When the assumed life "t" is less than the estimated life of a particular project, the project net operating incomes beyond the common assumed terminal point are discounted to the assumed terminal point. Thus, we have accrued an equivalent sum **F** after a period of **t** years. Next, the investments are discounted to time 0 at the company average investment opportunity rate. The AOEROR, then, is the interest rate that a bank would have to pay on the equivalent time 0 investment in order to accrue the sum **F** at the end of **t** years.

Example 3-12
Calculating AOEROR

The following NOI's are estimated to result from investments of $250 at time 0 and $616 at the end of year 1.

Year	Investment ($)	NOI ($)
0	250	0
1	616	1100
2		300
3		70

For this example let's assume a terminal point of ten years. To calculate the AOEROR, we need:

(1) The present worth of all investments discounted at the average investment opportunity rate. If i_{op} =10%.

$$I = 250 + \frac{616}{(1.10)^1} = \$810$$

(2) The future worth of all NOI's reinvested at i_{op}. (You may find a cash flow diagram helpful.)

$$F = (1100)(1.10)^9 + 300(1.10)^8 + 70(1.10)^7$$

$$F = \$3,373$$

$$AOEROR = \left(\frac{F}{I}\right)^{\frac{1}{t}} - 1 = 15.3\%$$

The equation in its simplest form may be expressed as:

$$I(1 + i)^t = F \tag{3-10}$$

where:

- t = assumed common terminal point in years
- I = present value of the investments discounted at i_{op}
- i = appreciation of equity rate of return
- F = the equivalent future sum of money that accrues when the NOI's are invested at the average investment opportunity rate until t years is reached.

3.4.6.2 *Concept*

Treasury growth is the fundamental concept underlying the AOEROR yardstick. The assumption that all future cash flows will be reinvested at the company's average investment opportunity rate assures that projects will be selected that will cause the treasury to grow as fast as possible. An equivalence or comparison for management is also established. Management can say: ''Am I willing to forego an amount ''**I**'' today, and receive an amount ''**F**'' at the end of year **t**.'' This is equivalent to putting the amount **I** in a bank that will pay an interest rate **i**, the appreciation of equity rate of return. At the end of **t** years you can withdraw the sum ''**F**'' from the bank.

One variation in the AOEROR assumes the project life rather than a common terminal point for all projects. The assumption of a common terminal point is important to make the AOEROR ranking of projects consistent with our discounted profit to investment ratio. This can best be seen by rearranging the AOEROR equation and solving for i.

$$i = \left(\frac{F}{I}\right)^{\frac{1}{t}} - 1 \tag{3-11}$$

Or, in the more complete form:

$$i = \left[\frac{\sum_{j=1}^{L} NOI_j\,(1 + i_{op})^{L-j}}{\sum_{j=0}^{L} \frac{INV_j}{(1 + i_{op})^j}}\right]^{\frac{1}{L}} - 1 \tag{3-12}$$

where:

- NOI_j = net operating income for period j
- INV_j = investment period j
- i_{op} = average investment opportunity rate
- i = AOEROR
- L = terminal point

For the case where ''L'' is common to all projects, the expression in brackets becomes equivalent to the discounted profit to investment ratio except that the net operating incomes are measured at a common point in the future rather than at time 0. Thus, when a common ''life'' is assumed for all projects the ranking using the AOEROR is equivalent to the ranking using the discounted profit to investment ratio when investments and NOI's are discounted at the average investment opportunity rate.

3.5 COMPARING ALTERNATIVES

In previous paragraphs, examples were provided to demonstrate how to use the various yardsticks when comparing investment alternatives. This section will consolidate the appropriate application of the various yardsticks. To do this we will compare and rank seven projects using the various yardsticks that we have discussed. We do this to demonstrate that the various yardsticks will not necessarily rank projects in a similar manner. Rather, it is important to understand the limitations inherent in each yardstick before a yardstick is used to make an investment decision.

The following table presents the cash flow analysis for the seven projects.

PROJECT CASH FLOW PROJECTIONS

	A		B		C		D	
YEAR	NOI ($)	INV ($)	NOI ($)	INV ($)	NOI ($)	INV ($)	NOI ($)	INV ($)
0	0	1100	0	500	0	1100	0	500
1	900		0		800		800	
2	200		0		300		300	
3	150		0		150		150	
4	75		1000		75		75	
5	75		900		75		75	
6	50		800		25		25	
7	50		300		25		25	
8	20		150		20		20	
9	20		75		0		0	
10	20		75		0		0	
	1560	1100	3300	500	1470	1100	1470	500

	E		F		G	
YEAR	NOI ($)	INV ($)	NOI ($)	INV ($)	NOI ($)	INV ($)
0	0	500	0	250	0	850
1	1100		1100	616	0	
2	300		300		0	
3	70		70		0	
4	0				500	
5	0				900	
6	0				800	
7	0				300	
8	0				150	
9	0				75	
10	0				75	
	1470	500	1470	866	2800	850

Using the methods developed in this chapter, several yardsticks were calculated for each of the projects. Three methods ignored the time value of money; five methods incorporated the time value of money. The results are presented in Table 3-1 and Table 3-2 which follow.

Table 3-1

RANKING PROJECTS

METHODS THAT IGNORE TIME VALUE OF MONEY

	Payout (Years)		UNDISCOUNTED NOI / INV		CUMULATIVE UNDISC. NCF($)	
RANK	PROJECT	VALUE	PROJECT	VALUE	PROJECT	VALUE
1	E	0.45	B	6.6	B	2800
2	F	0.53	H	3.29	H	1950
3	D	0.63	D	2.94	D	970
4	A	2.00	E	2.94	E	970
5	C	2.00	F	1.70	F	604
6	B	3.50	A	1.42	A	460
7	G	4.40	C	1.34	C	370

Table 3-2

RANKING PROJECTS

METHODS THAT INCORPORATE THE TIME VALUE OF MONEY

	NPV @ 10%		NPV @ 20%		DCFROR	
RANK	PROJECT	VALUE	PROJECT	VALUE	PROJECT	VALUE
1	B	$1478	B	$757	F	146.8%
2	E	800	E	666	E	146.6
3	G	787	D	548	D	100.4
4	D	722	F	402	B	44.2
5	F	491	G	166	G	24.0
6	A	173	A	-15	A	19.0
7	C	122	C	-52	C	16.7

	MODIFIED DCFROR		AOEROR @ i_{op} = 10%		NOI / INV @ i_{op} = 10%	
	PROJECT	VALUE	PROJECT	VALUE	PROJECT	VALUE
1	E	146.6%	B	26.2%	B	3.96
2	D	100.4	E	21.0	E	2.60
3	F	62.0	D	20.3	D	2.44
4	B	44.2	G	17.5	G	1.92
5	G	24.0	F	15.3	F	1.61
6	A	19.0	A	11.6	A	1.61
7	C	16.7	C	11.2	C	1.11

It should be obvious that consistent rankings are not obtained when using the various techniques. This is the case, as we have discussed, because the underlying assumptions for each method are different. The following section will discuss some of the specific project rankings.

- Project E ranks number 1 using payout and the modified DCFROR criteria. However, the undiscounted NOI to invest ment ratio ranks Project E fourth.
- Project B ranks number 1 using several techniques but only ranks number 6 using the payout concept.
- Project G ranks third when the NPV is discounted at 10% versus fifth when the NPV is discounted at 20%. The NCF's for Project G are delayed until the fourth year.
- Project F ranks number 1 if DCFROR is the primary yardstick. However, if a realistic assumption is made concerning the reinvestment of the cash flows, a significantly different ranking is obtained. For example, for i_{op} = 10%, the NPV method ranks Project F fifth.
- Project F has an investment subsequent to time 0. When using the conventional DCFROR method, the calculated DCFROR is 146.8% which results in a ranking of number 1. However, when subsequent investments are discounted at i_{op} rather than the DCFROR, the modified DCFROR is 62% which is significantly lower. Project F ranked third using the modified DCFROR technique.
- The AOEROR discounted at i_{op} and NOI/INV ratio also discounted at i_{op}, result in identical rankings. This will always be the case as long as the investment opportunity rate is used to discount the cash flows and the project AOEROR's are calculated for a common time line.

3.5.1 *Which Yardstick Is Best?*

After studying the underlying concepts and calculating several yardsticks, one should ask which method is best? In the final analysis, the best method should result in cost effective decision making. As you may expect, it is impossible to say in a given situation, which method will give the most cost effective decision. However, all things considered, there are certain methods which are better from a conceptual viewpoint.

3.5.1.1 *Single Projects*

When screening single projects, NPV at the investment opportunity rate (i_{op}) is easy to use and leads to the correct decision. If NPV is positive, the project is acceptable. If the investment opportunity rate is not known with certainty, plot the present value profile. If the NPV is positive for all possible i_{op}'s then the project is acceptable.

3.5.1.2 *Mutually Exclusive Projects*

Again NPV at i_{op} is the recommended yardstick. Pick the project with the highest NVP. If you are uncertain about i_{op}, plot the present value profile (PVP) for each project on the same graph. In fact, it's a good idea to plot the PVP so that the sensitivity of the decision to i_{op} can be determined.

3.5.1.3 *Capital Budgeting*

Assuming that the corporate goal is treasury growth, pick the projects that yield the greatest combined AOEROR. This sounds formidable, but a simple technique to do this is available. Simply rank the projects on the basis of NOI to investment ratio discounted at the investment opportunity rate. This assures that the projects are ranked the same as they would have been using AOEROR. Thus the rate of treasury growth is maximized. Similar results can be obtained by combining the various projects until the NPV discounted at i_{op} is the greatest.

References

Capen, E.C., Clapp, R.V, and Phelps, W.W., ''Growth Rate — A measure of Investment Efficiency and a challenge to current Economic Folklore'' paper presented at 48th Annual Meeting of Society of Petroleum Engineers, Las Vegas, Nevada, Sept. 30 - Oct. 3, 1973 (SPE Preprint No. 4613)

Phillips, Charles E., ''The Appreciation of Equity Concept and Its Relationship to Multiple Rates of Return,'' Journel of Petroleum Technology, February 1965, p. 159-163.

Roebuck, Field, ''Minimum-Interest Rate of Return Optimizes ROI'', Oil & Gas Journel, February 20, 1984, p.69-70.

PROBLEMS

3-1 The following revenues, costs, salvage value, and depreciation figures have been prepared for a centrifugal pump installation. How many years and/or parts of a year will it take for this investment to pay out? Assume receipts and payments occur evenly over a year.

	Year 0	1	2	3	4	5
Revenue		$ 500	$ 1000	$ 2000	$ 1000	$ 3000
Direct Costs		800	500	500	500	500
Depreciation		1000	500	250	250	0
Salvage Value						400
Investment	$ 2000					

3-2 Your boss tells you that accounting has generated the following numbers for a compressor installation, and he wants you to determine the after tax discounted net cash flow to investment ratio assuming mid-year discounting and an 18% per year discount rate (annual compounding). The first year is one year long. The required investment is $100,000 at the beginning of the project. What is the discounted profit to investment ratio?

Year	Gross Income	Direct Costs	Depreciation	Federal Income Tax
1	100,000	20,000	25,000	17,000
2	90,000	20,000	25,000	20,000
3	80,000	20,000	25,000	16,000
4	60,000	20,000	25,000	4,000

3-3 Calculate the modified DCFROR and standard DCFROR for the following proposed water flood installation to the nearest whole percent. Your average reinvestment opportunity rate is 12%. Assume year end discounting. All amounts are in thousands of dollars.

Year	Investment	Net Operating Income
0	800	—
1	0	900
2	500	600
3	0	400
4	0	200

3-4 You have a choice between the following two mutually exclusive projects. Assume cash is received at the end of the year. Enough money is available to do either project. Also assume that projects having an average return are available in any size.

Year	NCF $ E	NCF $ F
0	<10,000>	<11,000>
1	14,000	
2		
3		
4		21,000

a. Which project would you recommend if you have a reinvestment opportunity interest rate of 10%? What is the NPV at this rate?

b. Which project would you recommend if your reinvestment opportunity rate is 12%? What is the net present value at this rate?

c. What is the DCFROR for each project?

3-5 a. Assume that you deposit $12,500 in a bank paying 13% compounded annually. At the end of each of the following eight years you withdraw the sums indicated in the table. How much will you have in the bank at the end of the eighth year?

End of Year	Interest earned on balance in bank during Year	Balance in bank at end of year before Withdrawal	Withdrawal	Balance at end of year after Withdrawal
0	—	$12,500	—	—
1			$3800	
2			3100	
3			2500	
4			2200	
5			1800	
6			1700	
7			2300	
8			2275	

b. An investment of $12,500 for a new sucker rod string is estimated to generate the following incremental cash flows:

End of Year	
1	$3800
2	3100
3	2500
4	2200
5	1800
6	1700
7	2300
8	2275

What is the DCFROR for this investment?

3-6 Calculate the AOEROR based on the project life for the following proposed waterflood installation to the nearest whole percent. Your reinvestment opportunity rate is 12%. Assume end year discounting. All amounts are in thousands of dollars.

Year	Net Operating Income	INV
0	—	$ 800
1	$ 900	
2	600	500
3	400	
4	200	
5	200	
6	100	
	2400	

Discounted PV of NOI's @ 12% = $1,858

3-7 An offshore platform is used to dehydrate natural gas before it reaches the sales meter. Currently the gas does not meet pipeline specifications because of the water content of the gas. The contract specifies that the water content should *not* exceed 5 lb/MMSCF. However, a dew point check shows that consistently the water vapor content of the gas is 8 lb/MMSCF. Two options are available at the present time.

OPTION A:

Install larger dehydration facilities.

Investment = $185,000
Current salvage value of old dehydrator $60,000
Salvage value of new dehydrator
(end 15 years) $25,000
Operating costs $0.25 /MCF

OPTION B:

Take a reduction in the price. The transmission company will pay $3.50/MCF if the gas meets the specifications *and* $3.00/MCF if the gas does not meet the specifications. Operating costs are $0.25/MCF using the old dehydrator. The old dehydrator will *not* have a salvage value at the end of 15 years.

GAS PRODUCTION FORECAST

Time (Years)	BCF
1	10
2	10
3	10
4	9
5	8
6	7
7	6.5
8	6.5
9	6.0
10	6.0
11	5.5
12	5.5
13	5.0
14	4.5
15	4.0

OTHER INFORMATION:

It is expected that the gas will not meet specifications during years 1, 2 and 3. After year 3, there should not be a problem meeting pipeline specifications because of the lower flow rates and the capacity of the existing dehydration equipment. Your reinvestment opportunity rate is 15%.

3-8 What is the maximum amount that we can bid for an offshore lease if we have a 20% reinvestment opportunity rate. Assume annual componding. Do your analysis BFIT.

Time 0 =	BID
End of 1 year:	Drill exploratory well at cost of $2 MM.
End of 1.5 years:	Drill delineation well at a cost of $1.5 MM.
End of 2.0 years:	Drill delineation well at a cost of $1.5 MM.
End of 2.5 years:	Begin platform construction at a cost of $10 MM.
End of 3.0 years:	Continue platform construction at a cost of $10 MM.
End of 3.5 years:	Install pipeline — cost = $2 MM.
End of 4.0 years:	Set platform — cost = $1 MM.
End of 4.5 years:	Drill and complete wills = $20 MM. Commence production. See productiốn schedule. Assume end of period discounting for annual cash flow from production

PRODUCTION SCHEDULE

Time (Years)	Annual Production (Bbl)	NCF ($M)
5	1,300,000	32,500
6	1,300,000	32,500
7	1,100,000	27,50
8	667,000	16,675
9	404,000	10,100
10	245,000	6,125
11	150,000	3,750
12	90,000	2,250
13	50,000	1,250
14	33,000	825
15	20,000	500

End of 15 years: Salvage Value = abandonment costs

The present value of the production, discounted at 20%, at the beginning of year 5 is $ 98 MM.

3-9 Your company currently has several motor control panels for electric submersible pumps in stock. These panels have a maximum voltage rating of 1,000V. A new submersible pump installation is being planned in which a 100 horsepower downhole motor will be used. This pump can be powered with a motor requiring less than 1000 volts. However, if a panel having a larger voltage rating is purchased for $6,000 a motor with a lower amperage requirement can be used. The lower amperage motor will consume 1000 KW-HR's per day versus 1300 KW-HR's per day for the lower voltage motor (line losses are less). Should the larger voltage panel be purchased if power costs are $0.06/KW-HR, the average downtime is 5% and the project life is 15 years. Your time value of money is 20%. Assume any savings is generated at the end of each year.

3-10 You have a choice between the following two mutually exclusive projects. Assume cash is received at the end of the year.

Year	Net Cash Flow: Project 'A'	Net Cash Flow: Project 'B'
0	$ <25,000>	$ <25,000>
1	30,000	0
2	20,000	0
3	0	0
4	0	50,000
5	0	30,000

a. Plot the present value profile for each project on the same graph.

b. Which project would you recommend if your reinvestment opportunity interest rate is 15%? What is the NPV at this rate?

c. Which project would you recommend if your reinvestment opportunity interest rate is 20%? What is the NPV at this rate?

d. What is the DCFROR for each project?

3-11 Your company has had a good year and has extra money for projects. Your boss wants to propose an infill drilling project for a field that you are responsible for and wants you to perform calculations that will support a request to management for the $1 million required to drill the additional wells. The reservoir is rate insensitive which means that you do not need to be concerned with causing a decrease in reserves as a result of accelerated withdrawal. It is also extraordinarily homogeneous which means that you should not expect an increase in reserves as a result of the infill drilling. You have been given the following expected cash flows (assume they are received at the end of each year).

Year	Without Infill Drilling NCF ($Millions)	With Infill Drilling NCF ($Millions)
0	0	-1
1	5	10
2	4	4
3	3	1
4	2	0
5	1	0

a. Prepare a net present value profile of the *difference* between the two alternatives.

b. What are the minimum and maximum time values of money that favor the drilling?

c. If money is worth 15% to your company, should your boss make her request?

d. Suppose that problems are encountered which cause drilling costs to double. What are the minimum and maximum time values of money that favor the drilling?

3-12 a. Well 11-1 is currently producing at 3.0 BOPD. The motor on the pumping unit is run continuously. Power costs are $36/day. Smitty, the pumper, has 12 other nearby wells to operate and will therefore not be out of a job if this particular well is shut in. Incremental bookkeeping costs are also zero. Oil price is $25/Bbl (assume constant). We own a 100% working interest in the well. Landowner's and overriding royalties total 18% of gross revenue. Severance and advalorem taxes total 3% of gross revenue. It will cost $4,000 to pull the rods and tubing, plug, and abandon the well. This cost (assume constant) must be paid whether the well is abandoned now or at some time in the future. Assuming that no major repairs are required, calculate the economic limit flowrate (q_{el}) for this well.

b. Smitty has just come into the office and told us that the rods in well 11-1 have parted. Replacement rods will cost $200 (we are not replacing the whole string, only the damaged rods). Pulling and fishing costs are estimated to be $3,000. Assuming the production rate will decline at the rate of 0.1 BOPD per month (i.e. Q = 3.0 BOPD this month, 2.9 BOPD next month, etc.), should you repair the well? Money is worth 1% per month. Assume all cash flows occur at the end of each month.

3-13 A well is estimated to produce 400 BOPD from a depth of 5,000 feet. Given the following data, do you recommend a rod pump or an electric submersible pump installation? Your average investment opportunity rate is 15%. Assume costs occur at the end of each year.

	Investment	Oper Cost	Life	Salvage Value
Rod Pump	$42,000	$ 8,800/yr	10 yrs	$5,000
Subm Pump	$32,000	$20,440/yr	10 yrs	$5,000

3-14 Your first assignment as a production engineer is to determine whether you should replace an existing pump now or two years from now. Your boss has given you the following information:

Project Life: 5 years

Existing Pump: $0 Salvage Value — 2 year remaining life
Operating Costs are $8,000 per year.

New Pump: Initial Cost: $22,000
Operating Costs would be $4,000 per year.

CHOICE: Replace existing pump now with the new unit which is more efficient. Salvage value of the new pump if purchased now would be $2,000 at end of 5 years.

OR: Replace existing pump after 2 years when it has worn out. The salvage value of the new pump at the end of the project would be $10,000 since it would have been in service for only 3 years.

Assume all operating costs are paid at the *middle* of each year. The time value of money is 12% APR compounded annually.

3-15 You are considering either buying or leasing a new BMW 730i. Money is worth 1% per month to you and your options are shown below.

Buy Option:

Pay $1,500 down
Pay $1,400 per month for 3 years
Sell at end of three years for $30,000

(continued)

Lease Option:

Pay no money down
Pay $950 per month for 3 years
Give back to leasing company at the end of three years

Which option do you prefer? What would the resale value of the car at the end of 3 years have to be to make the two deals equivalent?

3-16 After finding your dream house, you need to settle the financing. Two banks are willing to loan you the $50,000 that you need. The terms, however, are not the same. Loan A is a 30 year note with an annual percentage rate (APR) of 14%. Payments are to be made at the end of each month. Loan A also requires 3 points (3% of the loan value) to be paid at the closing. Loan B has a 15% APR with monthly payments. Loan B is also for 30 years but does not require any points to be paid at the closing. You intend to keep the house forever.

a. If money has a time value to you of 15%, which loan do you prefer?

b. How many points can you pay on loan A before you become indifferent between loan A and B? Your time value of money is 15%.

3-17 An existing pumping unit has the capacity to pump 200 B/D. However the well is not pumped-off. A fluid level is shot and shows that there is approximately 1800 feet of fluid above the pump. Inflow performance calculations indicate that the well is capable of producing 600 BPD at current reservoir conditions.

The division manager wants to increase the production since the division is currently below its forecast. Two options are available. Both options are for the remaining life of the well. Assume annual cash flows and end-of-period discounting. Your reinvestment opportunity rate is 26%. Make a recommendation.

Plan A

Buy a used pumping unit capable of 400 BOPD. The cost would be $42,000. Operating costs would be $24/day. Salvage value at the end of six years when the well is abandoned is $20,000. The existing unit will be stored in the warehouse. Cost to abandon is $5,000.

Plan B

Do nothing. Salavage value of existing unit at the end of seven years when the well is abandoned is $10,000. The existing pumping unit has operating costs of $17/day. The abandonment cost is $5,000.

PRODUCTION FORECAST

	PLAN A		PLAN B	
Time (Years)	Annual Production (Bbl)	Revenue	Annual Production (Bbl)	Revenue
1	138,700	$3,467,500	73,000	$1,825,000
2	74,369	1,859,225	73,000	1,825,000
3	33,306	832,650	66,612	1,662,800
4	15,056	376,400	33,762	844,050
5	6,844	171,100	15,056	376,400
6	3,016	75,400	6,844	171,100
7			3,016	75,400

Table of Contents
Chapter 4
Reserve Estimation

4.1 Definition of Reserves 4-1
4.1.1 Classification of Reserves 4-2
4.1.1.1 ARPS 4-2
4.1.1.2 SEC 4-2
4.1.1.3 SPE 4-4
4.1.1.4 An Editorial On The Sec Definitions 4-4
4.2 Methods of Calculating Reserves 4-6
4.2.1 Analogy 4-6
4.2.2 Volumetrics 4-7
4.2.2.1 Sources of Data 4-7
4.2.2.2 Methodology 4-7
4.2.2.2.1 Isolated Wells 4-8
4.2.2.2.2 Developed Fields 4-8
4.2.3 Material Balance 4-9
4.2.3.1 History Matching 4-10
4.2.3.2 Future Predictions 4-12
4.2.4 Decline Curves 4-12
4.2.5 Reservoir Simulation 4-12
4.2.6 Summary of Methods 4-13

Chapter 4

Reserve Estimation

An oil property evaluation is typically divided into two parts. 1) the remaining reserves are estimated and 2) the production schedule and cash flows are estimated. The reserve estimates have become an important component of evaluations because of SEC disclosure requirements and the large amount of attention that was focused on the oil industry during the late 70's and early 80's.

Even though one would think that ''reserves'' would be easy to define and maybe even fairly easy to determine, that is not the case. First we discuss reserve definitions and then we very briefly discuss the way reserves are estimated.

4.1 DEFINITION OF RESERVES

Reserves are those quantities of natural resources remaining to be produced. This definition implies certain conditions. First, in order to be called reserves, oil and gas must be *physically* produceable. Second, the oil and gas must be *economically* producible. Third, since the resource has not been produced and is inaccessible, reserves cannot be measured — they can only be estimated. Fourth, since reserves are remaining, there is a time line associated with each reserve estimate.

Economics definitely plays a part in the definition of reserves. In order to be called reserves, the hydrocarbon must be economically produceable. For example, there are billions of barrels of oil in place in the oil shale belt which runs through western Colorado. It has been demonstrated that it is physically possible to extract a crude-oil-like substance from the rocks. The *reserves* of oil in this area, however, sum up to zero. The reason is that it is uneconomical to produce oil from shale with existing technology (Another oil company is about to demonstrate this again).

We are normally interested only in those reserves which remain to be produced from this time forward. This leads to the term *remaining reserves*. We must speak of remaining reserves as of such and such a date (usually the effective date of the study). When we set the time line at the discovery date of the field or well, we can speak of *original reserves*. This term includes all oil and gas which is expected to be recovered over the life of the field or well. The authors prefer the term *ultimate recovery* which means the same thing, but avoids the use of the word ''reserves''. Remembering that we are dealing with estimates, the terms should be known as *estimated remaining reserves (ERR)* and *estimated ultimate recovery (EUR)*. This emphasizes that they are estimates and not measured quantities.

Hydrocarbons which have been produced up to the effective date of the study are known as *cumulative recovery* (Cum — pronounced Cūm as in cumulative). Again, Cum has a time associated with it. The relationship between EUR, Cum, and ERR is as follows.

$$\text{EUR} = \text{ERR} + \text{Cum} \qquad (4\text{-}1)$$

In some cases the cumulative production may not be known as of the effective date since production data normally takes a month or two to get through the system. In this case it will be necessary to estimate the production from the last measured data to the effective date. We therefore may wind up with an *estimated cumulative recovery* as of the effective date.

Earlier it was mentioned that not all of the oil in place is recoverable. The term which relates EUR to original oil in place (N) is *recovery factor* (RF). In equation form:

$$EUR = N * RF \qquad (4\text{-}2)$$

We can combine equations 4-1 and 4-2 as follows:

$$ERR = N * RF - Cum \qquad (4\text{-}3)$$

Only the **Cum** term is measurable. All of the others are estimates. As we show later in the chapter there are some independent methods of estimating **ERR, N**, and **RF** which allow us to cross-check our answers.

4.1.1. *Classification Of Reserves*

Over the years, various people and various groups have attempted to define different categories of reserves.

4.1.1.1. *ARPS*

J.J. Arps published one of the more complete classifications in 1962. He classified reserves by four criteria:

1) Source of Reservoir Energy
   a) Primary
   b) Secondary
2) Degree of Proof
   a) Proved
   b) Probable
   c) Possible
3) Development Status (Proved Only)
   a) Developed
   b) Undeveloped
4) Producing Status (Proved Developed Only)
   a) Producing
   b) Nonproducing

Table 4-1 shows his classifications.

Since 1962, the American Petroleum Institute (API), the Society of Petroleum Engineers (SPE), and the Securities and Exchange Commission (SEC) have worked on the definition of Proved reserves.

4.1.1.2. *SEC*

The SEC is interested only in "proved" reserves, and does not classify them by producing mechanism or producing status. Therefore they only have two categories:

1) Proved Developed Reserves
2) Proved Undeveloped Reserves
   and, the total, Proved Reserves.

The following SEC definitions are taken from SEC Regulation 210.3-18:

"*Proved Oil and Gas Reserves*. Proved oil and gas reserves are the estimated quantities of crude oil, natural gas and natural gas liquids which geological and engineering data demonstrate with reasonable certainty to be recoverable in future years from known reservoirs under existing economic and operating conditions, i.e. prices and costs as of the date the estimate is made. Prices include consideration of changes in existing prices provided only by contractual arrangements, but not on escalations based upon future conditions.

(i) Reservoirs are considered proved if economic producibility is supported by either actual production or conclusive formation test. The area of a reservoir considered proved includes (A) that portion delineated by drilling and defined by gas-oil and/or oil-water contacts, if any, and (B) the immediately adjoining portions not yet drilled, but which can be reasonably judged as economically productive on the basis of available geological and engineering data. In the absence of information on fluid contacts, the lowest known structural occurrence of hydrocarbons controls the lower proved limit of the reservoir.

(ii) Reserves which can be produced economically through application of

Table 4-1
ARPS' CLASSIFICATION OF PETROLEUM RESERVES

Source of Reservoir Energy	Degree of Proof	Development Status	Producing Status
PRIMARY Reserves recoverable commercially at current prices and costs, by conventional methods and equipment, as a result of natural energy inherent in the reservoir.	PROVED Primary reserves which are considered proved to a high degree of certainty by actual production from the reservoir at commercial rates, or in certain cases by successful well test(s) in conjunction with favorable and reliable core analysis data or quantitative log interpretation. This indirect evidence is valid only if no substantial drainage has occurred since these well(s) were drilled or tested.	DEVELOPED Proved reserves recoverable through existing wells and with existing facilities.	PRODUCING Developed reserves to be produced by existing wells from completion interval(s) open to production.
			NONPRODUCING Developed reserves to be produced from existing wells but which are now behind the casing or at minor depths below the bottom of the hole. The cost of opening up such reserves should be relatively minor.
		UNDEVELOPED Proved reserves under undeveloped spacing units which are so close and so related to developed spacing units that they may be assumed with confidence to become commercially productive when drilled.	
	PROBABLE Primary reserves behind the casing of existing wells or within the known geological limits of a productive reservoir, which are inferred from limited evidence of commercially producible oil or gas, but where the evidence is insufficient to qualify under the "proved" definition.		
	POSSIBLE Primary reserves whose existence may be inferred from geological considerations but where available data will not support a higher classification.		
SECONDARY Reserves recoverable commercially at current prices and costs, in addition to the primary reserves, as a result of supplementing by artificial means the natural energy inherent in the reservoir; sometimes in conjunction with a change in the physical characteristics of the reservoir fluids.	PROVED Secondary reserves which are considered proved to a high degree of certainty by a successful pilot operation or by satisfactory performance of full scale secondary operation in the same reservoir, or in certain cases in a similar nearby reservoir producing from the same formation.	DEVELOPED Proved reserves, recoverable through existing wells from a reservoir where successful secondary operations are in progress.	PRODUCING Developed reserves to be produced by existing wells from that portion of a reservoir subjected to full scale secondary operations.
			NONPRODUCING Developed reserves to be produced by existing wells upon enlargement of existing secondary operations, provided the cost of such expension is relatively minor.
		UNDEVELOPED Proved reserves which may be assumed with confidence to be produced upon the installation of a secondary recovery project and/or by the drilling of additional wells.	
	PROBABLE Secondary reserves which are inferred from past production performance or core, log, or reservoir data, but where the reservoir itself has not been subjected to secondary operations.		
	POSSIBLE Secondary reserves from reservoirs which appear to be suited for secondary operations but where available data will not support a higher classification.		

improved recovery techniques (such as fluid injection) are included in the ''proved'' classification when successful testing by a pilot project, or the operation of an installed program in the reservoir, provides support for the engineering analysis on which the project or program was based.

(iii) Estimates of proved reserves do not include the following: (A) oil that may become available from known reservoirs but is classified separately as ''indicated additional reserves''; (B) crude oil, natural gas, and natural gas liquids, the recovery of which is subject to reasonable doubt because of uncertainty as to geology, reservoir characteristics, or economic factors; (C) crude oil, natural gas, and natural gas liquids, that may occur in undrilled prospects; and (D) crude oil, natural gas, and natural gas liquids, that may be recovered from oil shales, coal, gilsonite and other such sources.

Proved Developed Oil and Gas Reserves. Proved developed oil and gas reserves are reserves that can be expected to be recovered through existing wells with existing equipment and operating methods. Additional oil and gas expected to be obtained through the application of fluid injection or other improved recovery techniques for supplementing the natural forces and mechanisms of primary recovery should be included as ''proved developed reserves'' only after testing by a pilot project or after the operation of an installed program has confirmed through production response that increased recovery will be achieved.

Proved Undeveloped Reserves. Proved undeveloped oil and gas reserves are reserves that are expected to be recovered from new wells on undrilled acreage, or from existing wells where a relatively major expenditure is required for recompletion. Reserves on undrilled acreage shall be limited to those drilling units offsetting productive units that are reasonably certain of production when drilled. Proved reserves for other undrilled units can be claimed only where it can be demonstrated with certainty that there is continuity of production from the existing productive formation. Under no circumstances should estimates for proved undeveloped reserves be attributable to any acreage for which an application of fluid injection or other improved recovery technique is contemplated, unless such techniques have been proved effective by actual tests in the area and in the same reservoir.''

4.1.1.3. *SPE*

In 1981, the SPE board of directors approved a set of definitions for Proved Reserves. These differ so little from the SEC definitions that they will not be reprinted here. Readers interested in the SPE definitions can find them on page 2113 of the November 1981, Journal of Petroleum Technology.

4.1.1.4. *An Editorial On The SEC Definitions*

It is unfortunate that the term ''proved'' has been attached to reserves. They are never *proved* — they are always *estimated*! In fact, the only thing which can be said about reserve estimates with any certainty is: No matter what answer you come up with — it is wrong. The reserve estimator must learn to live with this. Hopefully the answer he gets is not too far off and hopefully he is evaluating enough properties that he is high on some and low on others. In that case the reserve estimator's savior — the law of compensating errors — will once again come to his rescue.

Let's look at some of the key phrases in the SEC definitions.

Proved Reserves

1) estimated quantities.
2) reasonably certain to be recoverable.
3) doesn't include products whose recovery is subject to reasonable doubt.
4) existing economic and operating conditions.
5) actual production or conclusive formation test in reservoir.
6) enhanced recovery must have pilot test or actual operation.

Proved Developed Reserves

1) existing wells with existing equipment.
2) enhanced recovery must have pilot test or have seen response.

Proved Undeveloped Reserves

1) new wells on offset acreage reasonably certain of production.
2) existing wells requiring major expenditure.

The definition of Proved Developed is pretty good. The name should be changed to ''Estimated'', however. Notice the difference in connotation between the following sentences:

1) XYZ Oil Company has Proved Developed Reserves of 397,216 STB.
2) XYZ Oil Company has Estimated Developed Reserves of 397,000 STB.

Sentence 1 sounds like a guarantee. Sentence 2 sounds like they're not sure. *Indeed, they are* **not** *sure* — in fact, they're probably not even close. So why make it sound as if reserves are like sheep which can be counted?

The inclusion of Proved Undeveloped Reserves in an SEC report is a disaster! Reserve Estimators would be much better off if they lived by the credo: THERE ARE NO PROVED UNDEVELOPED RESERVES.

Many, many times a reserve estimator will assign value to offset acreage. The operator then drills the well and, *surprise*, there is no sand. The company then gets to write down its proved (guaranteed?) reserves. The stockholders sue, the bank calls its loans, the board of directors resigns, and everybody points at the reserve estimator. One of the most notorious problems occurs with shut-in gas wells. Two or three gas wells are drilled in a low permeability formation and tested for 4 hours at 2 MMSCFPD (each). They are then shut-in awaiting the pipeline. A reserve report is prepared which shows the wells as proved developed (correctly, under SEC definitions) and shows all interior locations and one offset location in each direction as proved undeveloped. When the wells are put on production, the rates drop like a rock. Pretty soon they're making 200 MCFPD and it's uneconomical to drill the infills and offsets. Cut the reserves, the stockholders sue, etc. This generic and ''far fetched'' example is based on an actual case in which the reserve estimator got sued.

Of course, the way to avoid all of this grief is to not book the proved undeveloped reserves in the first place. *Right*! The pressure put on ''independent'' reserve estimators and in-house reserve estimators is incredible. The president of a Small Independent Oil Company and the division manager of a Large Major Oil Company want the reserves up. One so his insider stock goes from 1¢ per share to $2 per share, and the other so that he'll get more than his allocated 10% raise or show that his exploration project really is a success. It is extremely difficult for the reserve estimator to resist this pressure. One of the more sophisticated ways of pushing the reserve estimates up is for the client (or manager) to ''review'' the report with the estimator. On those wells where the client thinks the estimator is high, the client smoothly turns the page. On those wells where the estimator is low, out come the thumbscrews. This perverts the law of compensating errors.

What neither the client nor the manager realizes is that the **truth will eventually be known**. The small oil company stockholder cannot trade his ''insider'' stock for two years and even then he won't be able to unload all of it at once. By that time investors usually will have figured out what is going on and his stock will be worthless. The manager intends to be a VP someday. After he's out of the division how is he going to keep inflated reserve estimates quiet? His replacement sure isn't going to get stuck with them.

It is necessary to take a long term view. If the ''proved undeveloped'' reserves are indeed there, they will still be there next year. So drill them up and book them as proved developed.

The SEC could help by not allowing proved undeveloped reserves. In fact the authors suggest that reserves be divided into two categories:

Category I — Reserves from wells which have produced for at least 6 full, continuous months.

Category II — Reserves from wells which have been drilled, completed, and *placed on production*, but have less than 6 months history.

This would cut down considerably on the abuses which took place during the last few years. If the SEC wanted companies to disclose more information they could allow a separate section discussing "possible" or "speculative" or "indicated additional" reserves with the appropriate caveats.

Now that we have defined and editorialized about reserves let's examine the various methods to calculate them and discuss the advantages and disadvantages of each method.

4.2. METHODS OF CALCULATING RESERVES

There are four traditional methods of estimating reserves and one relative newcomer. They are:

1) Analogy
2) Volumetrics
3) Material Balance
4) Decline Curves
5) Reservoir Simulation (the newcomer)

Reservoir simulation is essentially an extension of the material balance technique. It is also quite expensive and is not routinely used for reserve estimation so it will be discussed separately. The remarks which follow are related to the four major methods of estimating reserves.

Each of the four has advantages and disadvantages and each can be applied independently. That is, each method requires different data than the others to arrive at the same result — recoverable reserves. Because of this independence, the different techniques can be used to crosscheck the answers.

4.2.1. *Analogy*

Analogy is one of the most often used methods of reserve estimation. It can be as unsophisticated as "the well across the road made 100,000 Bbls, this one ought to, also" or as sophisticated as deriving relative permeability curves from offset production peformance. Most people tend to think of analogy as the unsophisticated approach, but we use it almost every time we do reserves. Some of the ways it is used are:

Recovery Factor — We almost always look to offset acreage for recovery factors (RF). In fact, it is worthwhile to do reservoir studies on fields close to abandonment so that good values of recovery factor are available for use in other fields.

Bbls per acre foot — Barrels per acre foot (BAF) is similar to recovery factor, but lumps more of the terms in the volumetric equation together. The equation for BAF is as follows. (Terms are defined in the next section).

$$\text{BAF} = (7758)\ \frac{\phi\ (1 - Sw_i)}{B_{oi}}\ \text{RF} \qquad (4\text{-}4)$$

In order to use BAF you must assume that the product

$$\frac{\phi\ (1 - Sw_i)}{B_{oi}}\ \text{RF}$$

is the same between your well and the analogous well. This is most useful when you are running the economics on a yet-to-be-drilled exploratory well.

EUR — Comparing EUR's is done during the exploratory phase. It is also quite useful as a comparison when calculating proved developed reserves. If your proved developed reserves are twice as high as the EUR from the nearest 100 wells, you had better check your assumptions. You may indeed have the best well in the county, but you'd better have some good supporting evidence.

Care should be taken to make sure the field or well being used for analogy is indeed analogous. A Red River dolomite with volatile crude will probably not be at all similar to a Muddy sand with black oil. This point should be obvious, but it is often violated, especially by using BAF.

4.2.2. *Volumetrics*

A volumetric estimate of oil-in-place is one of the most important methods of determining reserves.

The method can be very simple, requiring only well logs and some estimated parameters. Since it requires only logs, a volumetric analysis can be done before the well is completed. Indeed, it is often done at the drillsite to decide whether or not to complete a well. The analysis can also be quite complex, requiring the use of core data, special core analyses, suites of many types of logs and apparently sophisticated computer programs. No matter how simple or complex the analysis, the fundamental equation for oil reservoirs is:

$$EUR = \frac{7758\ Ah\phi\ (1 - S_{wi})}{B_{oi}}\ RF \qquad (4\text{-}5)$$

where

- EUR = estimated ultimate recovery, STB
- A = drainage area, acres
- h = average height of formation, ft.
- ϕ = average porosity, decimal
- S_{w_i} = average initial water saturation, decimal
- B_{o_i} = initial oil formation volume factor, RB/STB
- RF = recovery factor, decimal
- 7758 = conversion factor, Bbls/Acre-ft.

The derivation of the equation is simple. The product **Ah** is the bulk volume of the reservoir. The pore volume in the reservoir is the bulk volume times the porosity ϕ. **Sw_i** represents the fraction of the pore volume initially occupied by water so **(1** − Sw_i) is the fraction initially occupied by oil. **Bo_i** converts reservoir bbls to stock tank bbls. The original oil-in-place (**N**) is:

$$N = \frac{7758\ Ah\phi\ (1 - S_{wi})}{B_{oi}} \qquad (4\text{-}6)$$

The EUR is, of course,

$$EUR = N * RF \qquad (4\text{-}2)$$

For gas, the equation becomes

$$EUR = \frac{43560\ Ah\phi\ (1 - S_{wi})}{B_{gi}}\ RF \qquad (4\text{-}7)$$

where

- EUR = estimated ultimate recovery, SCF
- B_{g_i} = initial gas formation volume factor, RCF/SCF
- 43560 = the number of square feet in an acre.

4.2.2.1. *Sources of Data*

A is estimated or, in a developed field, is based on the spacing. This often is one of the terms which is subject to large errors.

h is determined from well logs or cores.

ϕ is calculated from well logs or cores.

Sw_i is calculated from well logs.

Bo_i is determined from laboratory measurements of fluid samples or from correlations.

RF is usually determined from analogy or from correlations. This is the other term which is often subject to large errors.

4.2.2.2. *Methodology*

There are two ways of getting a volumetric estimate of oil-in-place. One method is applicable to isolated wells and the other to developed areas.

4.2.2.2.1 *Isolated Wells*

h, ϕ, and $\mathbf{Sw_i}$ are determined from well logs and the values at the well are assumed to apply throughout the entire drainage area. If we assume a sampling radius of 5 feet and a drainage area of 40 acres we have sampled about .004% of the drainage area. Luckily, nature is surprisingly uniform and we get away with this gross approximation (usually). On an isolated well, **A** is just a guess (excuse me, estimate). Look for nearby developed fields in the same formation. This will usually give a pretty good idea of the spacing that will be adopted. With formations that change radically over short distances (such as eastern Colorado D-Sands), **A** is not only a guess, but a wild guess. **RF** and $\mathbf{Bo_i}$ are determined as discussed above. A note on the calculation of average ϕ and average $\mathbf{Sw_i}$ is in order. ϕ is defined as the ratio of pore volume to bulk volume. If the reservoir is being modeled as a layered system, the equation for average porosity is

$$\overline{\phi} = \frac{\sum_{i=1}^{n} A_i h_i \phi_i}{\sum_{i=1}^{n} A_i h_i} \qquad (4\text{-}8)$$

where

A_i = Area for layer i
h_i = height for layer i
ϕ_i = porosity for layer i
n = the number of layers in the well.

Normally it is assumed that each layer has the same drainage area so the $\mathbf{A_i}$ terms all cancel. **Sw** is defined as the ratio of the volume of water to the pore volume. The equation for average water saturation is

$$\overline{Sw} = \frac{\sum_{i=1}^{n} A_i h_i \phi_i Sw_i}{\sum_{i=1}^{n} A_i h_i \phi_i} \qquad (4\text{-}9)$$

where

Sw_i = water saturation for layer i

Notice water saturation is *pore volume* weighted, not bulk volume weighted. Again the $\mathbf{A_i}$**'s** usually are assumed to cancel.

4.2.2.2.2 *Developed Fields*

In a developed area it is not necessary to assume that **h**, ϕ, and **Sw** are constant throughout the spacing pattern. The recommended method of calculating oil-in-place is as follows:

1) Calculate the product $\mathbf{h\phi(1\text{-}Sw_i)}$ for each layer in each well.

2) For each well sum the $\mathbf{h\phi(1\text{-}Sw_i)}$ values for all of the layers in that well.

3) Plot the sum for each well on a map of the field.

4) Contour the values. Often there will be very little control as to the location of the zero contour. Use geological judgement.

5) Planimeter the area enclosed by each contour.

6) Using cartesian coordinate paper plot the area enclosed by each contour on the X-axis and the value of the contour on the Y-axis. This results in a graph such as Figure 4-1.

7) Planimeter the area under the curve. This yields a value of $\mathbf{A\,h\,\phi\,(1\text{-}Sw_i)}$.

8) Multiply by 7758 and divide by $\mathbf{Bo_i}$ to get oil-in-place. Again, **EUR = N * RF**.

The contour map is referred to as an "isovol" map. This method is preferred to using average **h,** ϕ, and **Sw** values because it takes into account their variation across the field. Steps 6 and 7 could be replaced by the trapezoidal or pyramidal rule, but it is felt that the graph is useful in itself. It gives a picture of how the hydrocarbon pore volume is areally distributed. It may even be possible, after many fields are mapped this way, to be able to relate different shapes to different geological environments.

The above method works well to determine the total oil-in-place **(OIP)** in the reservoir, but we are usually interested in individual wells. Separating the total OIP into individual wells' OIP can sometimes be exceedingly difficult. In the case where we can assume that permeability is constant throughout the reservoir, single phase flow exists, and the compressibility is constant, it is quite easy to "split" the reservoir volume. Once pseudo-steady-state flow (see Chapter 5) has been achieved throughout the reservoir (often within a month), each well's drainage volume is proportional to its rate. In equation form:

$$V_i = \frac{q_i}{q_T} V_T \qquad (4\text{-}10)$$

where

V_i = drainage volume (**Ah**$\boldsymbol{\phi}$**(1-Sw$_i$)**) for well **i**, Acre ft.

V_T = total drainage volume of field, Acre ft.

q_i = production rate of well **i**.

q_T = total production rate for the field.

Through the use of a recovery factor, each well's recoverable reserves can now be determined.

Volumetric determination of oil-in-place is the keystone of reserve estimation. The required data is almost always available, it is reasonably quick to do, and it usually yields surprisingly good answers.

4.2.3. *Material Balance*

The material balance technique of determining original oil-in-place is based on the law of conservation of mass.

$$\text{mass produced} = \text{mass originally in place} - \text{mass currently in place} \qquad (4\text{-}11)$$

FIGURE 4-1

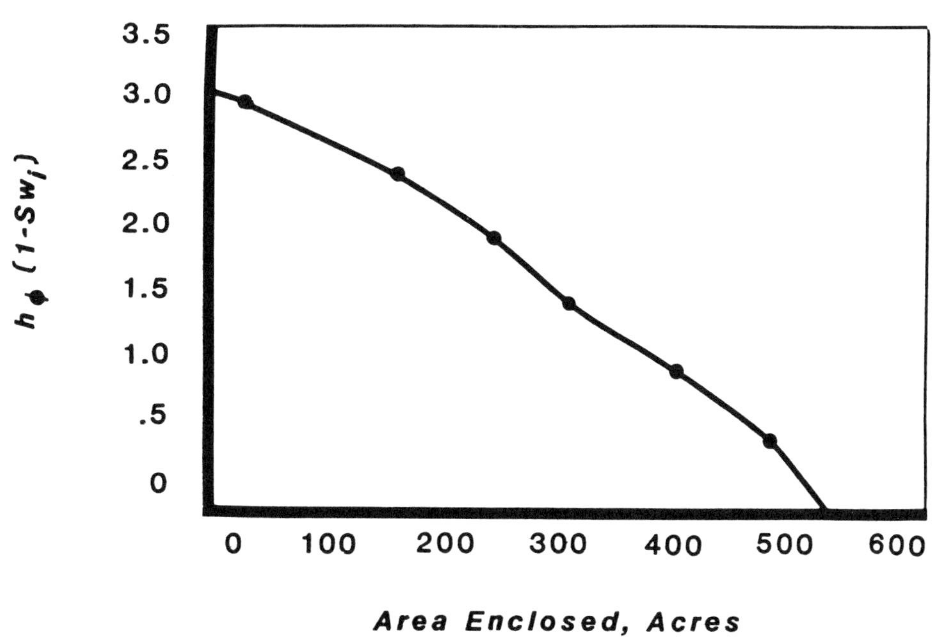

This equation can be applied to each component (C_1, C_2, etc.) of the hydrocarbon accumulation or the accumulation can be treated as either a single component (usually gas) or two components (oil and gas). Since we measure volume, not mass, the equation is usually expressed in terms of volume.

$$\text{Volume produced} = \text{Volume originally in place} - \text{Volume remaining} \quad (4\text{-}12)$$

As long as the density (at standard conditions) of the hydrocarbon being produced does not change, the conversion from mass to volume is acceptable. This is usually true in gas reservoirs and so-called ''black'' oil reservoirs. A black oil reservoir might be defined as a reservoir where the stock tank density of the crude does not change with time — no matter what color it is. Volatile oil reservoirs and gas condensate reservoirs, however, require special methods beyond the scope of this book.

The application of the material balance can be broken into two distinct phases — the determination of original oil and gas in-place (history matching) and the prediction of future performance.

4.2.3.1. *History Matching*

After the field has produced for some time, the amount of original oil-in-place can be determined. The material balance technique treats the reservoir as one single tank. In order to use this technique, it is necessary to know the amount of oil, gas and water produced, the average reservoir pressure (preferably at 6 or more different times), the pressure-volume-temperature relationships of the fluids, and the pore volume compressibility of the rock.

The complete material balance equation is given in *Petroleum Reservoir Engineering* by Amyx, Bass, and Whiting, page 566. It is rather awesome, but reduces to a much simpler equation for solution-gas-drive reservoirs with no water influx. Under those conditions the equation becomes

$$N = \frac{Np\left[B_t + (R_c - R_{si})\frac{Bg}{5.615}\right]}{B_t - B_{ti}} \quad (4\text{-}13)$$

where

N = original oil-in-place, STB.

Np = cumulative oil production, STB.

B_t = total formation volume factor at pressure associated with cumulative production of N_p, volume at reservoir conditions/volume at surface conditions, (RB/STB).

$$B_t = Bo + (R_{si} - R_s)\frac{Bg}{5.615}$$

B_{ti} = initial total formation volume factor.

R_c = cumulative gas-oil ratio, SCF/STB. $R_c = Gp/Np$

R_{si} = initial solution gas oil ratio, SCF/STB.

Bg = gas formation volume factor, RCF/SCF.

Bo = oil formation volume factor, RB/STB.

Rs = solution gas oil ratio at pressure associated with N_p, SCF/STB.

Np and $\mathbf{R_c}$ are measured. $\mathbf{B_{ti}}$ and $\mathbf{R_{si}}$ are known from fluid data and original reservoir pressure. **Bg** and $\mathbf{B_t}$ are determined from the PVT data at the reservoir pressure. With this data we can calculate a value of **N** for each known value of reservoir pressure. A graph of **N** vs. **Np** is plotted as Figure 4-2.

FIGURE 4-2

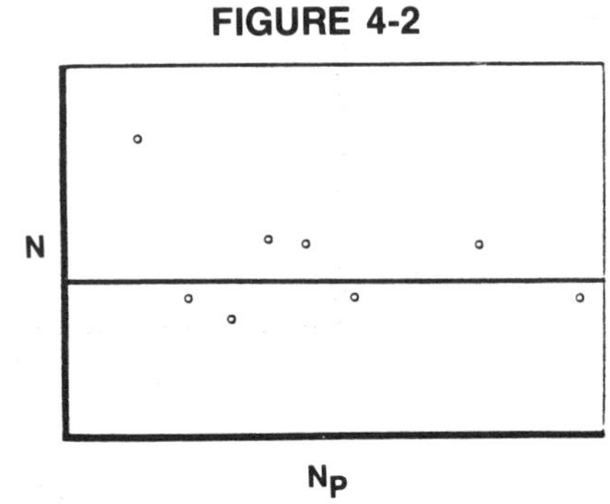

If the data points are randomly scattered and a horizontal straight line can be drawn through them, we can estimate the value of **N** and feel reasonably certain that no outside energy is being added to the system.

If there is water influx or a gas cap for which we have not accounted, the graph will appear as Figure 4-3.

FIGURE 4-3

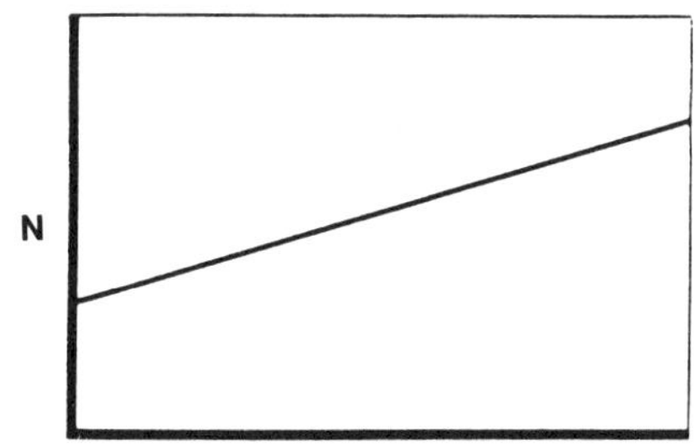

If you are not accounting for all of the voidage (offset wells that should be included but aren't) the graph will look like Figure 4-4.

FIGURE 4-4

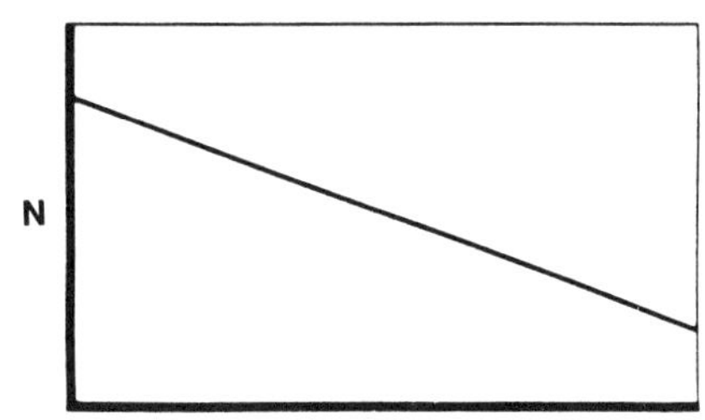

The material balance is extremely sensitive above the bubble point. Small variations in pressure can cause large changes in the calculated value of **N**. Below the bubble point the method becomes more stable because the gas compressibility is relatively large.

Many times the material balance method is hard to apply because of the lack of reliable pressure data. This is unfortunate because the method calculates the oil-in-place independent of its areal distribution. When material balance results are compared to volumetrics, it is possible to discover field extensions.

Material balance techniques are more routinely applied to gas reservoirs. The pressure data is more readily available because of regulatory requirements. When applied to a gas reservoir, the material balance equation is usually expressed as:

$$\frac{P}{Z} = \frac{P_i}{Z_i}\left(1 - \frac{Gp}{G}\right) \qquad (4\text{-}14)$$

where

P = "current" reservoir pressure, psia.

Z = "current" gas deviation factor.

P_i = initial reservoir pressure psia.

Z_i = initial gas deviation factor.

Gp = cumulative gas produced, MMSCF.

G = original gas in place, MMSCF.

This is the equation of a straight line with **P/Z** as the dependent variable and **Gp** as the independent variable. It plots as shown on Figure 4-5.

FIGURE 4-5

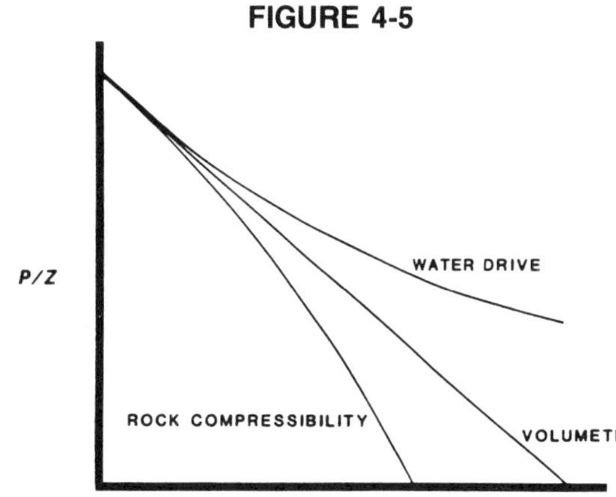

For a volumetric reservoir, **P/Z** plots as a straight line with an intercept on the x-axis at **G** (the original gas-in-place). Water influx will cause the curve to flatten. The other curve shows the shape of the curve when the pore volume compressibility is approximately the same as the initial gas compressibility. This can occur in reservoirs with an initial pressure of 10,000 psi or more. At high pressures the rock expansion can account for 30% to 50% of the energy in the system. Since the **P/Z** plot does not account for this energy source, extrapolating the initial data points can result in gas-in-place values which may be 50% to 100% too high. It is extremely difficult to tell that this is occurring from the initial data points since they plot almost exactly as a straight line. Be wary of **P/Z** plots on reservoirs with initial pressures above 10,000 psi.

4.2.3.2. *Future Predictions*

Once original oil-in-place has been determined, the material balance may be used to calculate future performance. Usually this takes the form of a **pressure** vs. **Np** plot and a **GOR** vs. **Np** plot, but, if absolute permeability is known, the results can be converted to a time basis. In order to calculate future performance, relative permeability curves must be available. It is often difficult to obtain these since they come from special core analysis. The future performance of gas reservoirs can be predicted by assuming that the P/Z plot will continue as a straight line (when appropriate).

A prediction of future performance gives us an estimate of recovery factor. This recovery factor can then be checked against other fields in the area to see if it is reasonable. One of the major problems with oil material balance predictions is that the hardest piece of data to obtain — relative permeability data — is the most important parameter. A slight change in relative permeability can cause a dramatic change in recovery.

4.2.4. *Decline Curves*

Decline curves are plots of producing rate vs. time. They are usually plotted on semi log paper and extrapolated to give an estimate of production vs. time. They do not require any assumptions about **A, h, ϕ, Sw** or **RF**. Since the only data requirement is production, the data is readily available. Decline curves are usually easy to analyze and often yield good answers. Their major disadvantage is their apparent simplicity. Because decline curve analysis is so deceptively simple and yields answers on a time basis, it is the most commonly applied method of reserve estimation. However, decline curves are not simple curves. In fact, they should be considered more difficult to analyze than pressure-transient tests. You might think of them as rate-transient tests. The next chapter is devoted entirely to decline curve analysis so it will not be discussed here. Just realize that they are not as easy as they look.

4.2.5. *Reservoir Simulation*

Although reservoir simulation is not used primarily for reserve analysis, it sometimes creates reserve estimates as a by-product. For that reason it will be briefly discussed.

Rather than represent the reservoir as one single cell as the material balance method does, a reservoir simulator represents the reservoir as a number of cells, all interconnected. These cells can be geometrically arranged in one, two, or three dimensions and can be as small or as large as desired. The fluid flow equations are used to calculate flow between cells.

A simulator requires large quantities of input data and produces prodigious amounts of output. For every cell in the model (typically 100's to 1,000's) the following information must be supplied:

Permeability (x, y, and z)
Porosity
Thickness
Elevation
Initial Pressure
Initial Saturations (So_i, Sw_i Sg_i)

Fluid characteristics can be assigned by "region" or a group of cells. The data which must be supplied for each region is:

Formation Volume Factors vs Pressure (oil, gas, water)
Viscosities vs Pressure (oil, gas, water)
Solution Gas-Oil Ratio vs Pressure
Densities (oil, gas, water)

Rock properties can also be assigned by region. They include:

Relative Permeability (oil, gas, water)
Capillary Pressures (oil-water, gas-oil)
Pore Volume Compressibility

Wells must be specified. The data necessary for each well includes:

Location
Producing Interval
Production History (oil, water, gas)
Pressure History
Productivity Index

Once all of this data is collected and put into the model, a history match is performed. During this phase, the above parameters are adjusted to match the pressure-production history of the wells. Hopefully, geological and engineering judgement is applied so that only reasonable values are used. This is not always the case. There is no such thing as a unique match and sometimes physically unreasonable models can match past history perfectly.

Once the history has been matched, the prediction phase begins. If we have a good history match with reasonable values for the parameters, we then feel better about our prediction run. There is no guarantee that, even with a good match, the prediction run is close to what will happen.

A reservoir simulator can be a very powerful tool in the hands of a competent engineer. It can also be used to create totally spurious results that look just as impressive as correct ones. We have seen one case where a field not yet on production was simulated. A certain number of barrels were predicted to be produceable. As more wells were drilled it became necessary to reduce the size of the reservoir (as you might expect, the actual sand thickness was less than predicted). In order to get the same number of barrels out of the field, the engineers fiddled with the relative permeability curves and increased the recovery factor! Admittedly, they didn't have any compelling basis for either set of relative permeability curves, but the approach is backwards. Don't work from the ''required'' answer to the model.

When looking at the results of a simulator run, remember that they don't necessarily tell the truth. A Reservoir Simulator is a tool, not an oracle.

4.2.6. *Summary Of Methods*

The following table lists the data requirements, advantages, and disadvantages for each of the methods of calculating reserves.

Table 4-2

Reserve Estimation Methods

1. ANALOGY

Requirements:	A field or well which is expected to perform similarly.
Advantages:	Fast, cheap, can be done before drilling.
Disadvantages:	Accuracy (Apples and Oranges).

2. VOLUMETRICS

Requirements:	A well. Logs and/or core. Estimate of drainage area, recovery factor (analogy), fluid properties (minor).
Advantages:	Minimal information. Can be done early in the life. Relatively fast.
Disadvantages:	Requires assumptions (area, recovery factor) which may not be true. May have gross errors.

3. DECLINE CURVES

Requirements:	Production history (only).

Advantages: No assumptions about size, type or other properties of reservoir. Need only production history. Fast, cheap. Very accurate under certain circumstances. Results in production versus time prediction.

Disadvantages: Well must be producing under "constant" conditions. Need at least 6 months history (better 2-10 years). Ambiguous (does not necessarily give unique answer). Cannot be used under changing well conditions. Not applicable to all reservoirs.

4. *MATERIAL BALANCE*

Requirements: Pressures, production history, fluid properties, rock properties (relative permeability required for prediction).

Advantages: No assumptions necessary for areal extent, thickness, recovery factor (at least not explicitly). Low sensitivity to porosity, water saturation. Can be used to calculate oil-in-place, gas-in-place, recoverable reserves (and therefore recovery factor), water influx, gas cap size.

Disadvantages: Pressures not usually available. Predictions are very sensitive to relative permeabilities. Requires more information than 1, 2, or 3.

5. *RESERVOIR SIMULATION*

Requirements: For each cell: permeability, porosity, thickness, elevation, saturation (initial), initial pressure, rock compressibility.

For each well: location, producing interval, production rates versus time, pressures versus time.

For each rock type: relative permeability of each phase, capillary pressures.

For each fluid type: formation volume factors, viscosity, gas solubility, density.

Reservoir description: faults, pinchouts, aquifers, layering.

Advantages: Ability to handle different rock and fluid properties in different areas of the reservoir. Can predict production from individual wells. Once history match is obtained, can study effects of different producing schemes. Input data requirements force close analysis of reservoir.

Disadvantages: Cost, time required to do study, amount of input data, non-unique match. Assumptions made to get match may not be true in prediction runs. People tend to believe the answers.

Notice that each method requires different data and goes about finding reserves in a different way. This creates exciting possibilities. If the material balance and decline curves say there is more oil-in-place than the volumetrics, then there are probably undrilled locations. By comparing the results from the various methods, much can be learned about the reservoir.

It seems really sad to cover the most important concept of reserve estimation in one small paragraph as we have just done. Let's expand upon the idea a little bit. As can be seen from the previous table, each method independently arrives at the same answer (ERR). Each method has different strengths, weaknesses and assumptions. By comparing the results, one can often isolate the faulty assumption and form a better picture of the reservoir. The classic example is the comparison of volumetrics to decline curve analysis. Calculate EUR (Cum + ERR) from decline curves and N (Original oil-in-place) from volumetrics. Divide EUR by N to get RF. From analogy, determine whether or not the RF is reasonable. If RF is high either the volumetrics are too low or the decline curve is not dropping as fast as it should. Being forced to reconcile the differences and develop a consistent picture of the reservoir is one of the most important steps of an oil property evaluation.

Table of Contents
Chapter 5
Decline Curve Analysis

5.1 Decline Curve Equations 5-2
5.1.1 Exponential Decline 5-2
5.1.1.1 Nominal and Effective Decline Rates 5-2
5.1.1.2 Cumulative Production 5-3
5.1.2 Hyperbolic Decline 5-4
5.1.2.1 Nominal and Effective Decline Rates 5-5
5.1.2.2 Cumulative Production 5-9
5.2 Theoretical Foundation of Decline Curves 5-10
5.2.1 History 5-11
5.2.2 Assumptions Required in Derivation 5-12
5.2.3 Equations 5-12
5.2.3.1 Rate-Time 5-12
5.2.3.2 Original Oil-in-Place 5-13
5.2.3.3 Permeability and Skin Factor 5-14
5.2.3.4 Shape Factor 5-14
5.2.3.5 Recovery Factor 5-14
5.3 Decline Curve Analysis 5-16
5.3.1 General 5-16
5.3.1.1 Plotting the Data 5-16
5.3.1.2 Correcting for Downtime 5-17
5.3.1.3 Multiple Wells on a Lease 5-17
5.3.1.4 Wild Fluctuations 5-18
5.3.2 Exponential Decline 5-18
5.3.2.1 Rate-Time Plots 5-18
5.3.2.2 Rate-Cum Plots 5-18
5.3.3 Hyperbolic Decline 5-19
5.3.3.1 Fetkovich Type Curves 5-19
5.3.3.1.1 Analytical Region-find kh/μ, r_e/r_w, ϕc_t 5-27
5.3.3.1.2 Empirical Region—find q_i, d_i, b 5-32
5.3.3.1.3 Moving qi and Di 5-32
5.3.3.1.4 Shifting the Curve 5-33
5.3.3.2 Trial and Error 5-33
5.3.3.2.1 Guessing "b" 5-34
5.3.3.3 Finding Di from a Tangent 5-37
5.3.3.4 Special Graph Paper 5-37
5.3.3.5 Ranges of "b" 5-37
5.3.3.6 Other Methods 5-40
5.4 Secondary Streams 5-40
5.4.1 Projecting Gas Production from an Oil Well 5-40
5.4.1.1 Constant GOR 5-40
5.4.1.2 Log Gp vs Log Np 5-40
5.4.2 Water-Oil Ratio vs Cumulative Oil 5-41
5.4.3 Projecting Condensate 5-43
5.4.3.1 Constant Yield 5-43
5.4.3.2 Hyperbolically Declining Liquid Yield 5-43
5.4.3.3 Log Np vs Log Gp 5-45
5.5 Summary 5-45

Chapter 5

Decline Curve Analysis

> ''Oil wells usually reach their maximum daily output shortly after they are completed. From that time they decline in production, the rapidity of decline depending on the output of the wells and on other factors governing their productivity. The production curve of a well shows the amount of oil produced per unit of time for several consecutive periods; if the conditions affecting the rate of production are not changed by outside influences, the curve will be fairly regular, and, if projected, will furnish useful knowledge as to the future production of the well. By the aid of this knowledge the value of a property may be judged, and proper depletion and depreciation charges may be made on the books of the operating company. Certain difficulties peculiar to oil production are encountered when estimating the future production of oil wells, and the oil producer has been more uncertain regarding the future production of his property than the producer of almost any other mineral.
>
> In recent years much attention has been devoted to this problem, and considerable progress has been made in the development of methods for estimating future production. As valuable as these methods are, to the producer who knows how to use them, they still leave much to be desired in accuracy and ease of application. Methods are needed that are more dependable, that will furnish closer estimates from data obtainable early in the life of the property, that show the limitation of the estimates, and that may be quickly and easily applied.''

The preceding quote, taken from the opening paragraph of an article by J. O. Lewis and Carl H. Beal which was presented to the AIME in *1918*, shows how, in spite of tremendous technological improvements, things have changed little in the last 60 years. Production curves are still useful for forecasting future production; there is still a large amount of uncertainty involved in that forecast; and, although much progress has been made in recent years, we still need to develop methods which are more dependable, and that may be quickly and easily applied.

This chapter discusses the latest developments in decline curve analysis, shows how to apply decline curves, and discusses the theoretical background of decline curves.

There is nothing more important than the production forecast in determining the value of oil and gas wells. The only other item which approaches the yearly estimate of production in importance is the price forecast. The late 70's and early 80's have seen an amazing roller coaster ride in prices that is probably not yet finished. But the fact remains, the resource must be there before anything else has meaning.

Decline curves are the most common means of forecasting production. As mentioned in the last chapter, decline curves have many advantages—they use data which is easy to obtain, they're easy to plot, they yield results on a time basis, and they're deceptively easy to analyze. Decline curves are also one of the oldest methods of predicting reserves.

Decline curves, as used today, are simply a plot of production rate versus time on semilog, log-log, or specially scaled paper. The most common plot is semilog. When the logarithm of producing rate is plotted versus linear time, a straight line often results. This phenomenon is referred to as "exponential decline" and is similar to the decay of a radioactive element. Exponential decline is also referred to as constant percentage decline because of terminology used in the early 1900's. Occasionally, someone will state that exponential decline and constant percentage decline are different. This is not true; they are synonyms for decline curves which plot as a straight line on semi-log paper.

Often the data will not plot as a straight line on semi-log paper, but instead will "curve up" or be concave upwards. This situation, in which the decline rate continuously decreases with time, can usually be modeled with a hyperbolic equation. In cases of this type, the well is said to be experiencing "hyperbolic decline." A special case of hyperbolic decline is known as "harmonic decline."

5.1 DECLINE CURVE EQUATIONS

5.1.1 *Exponential Decline*

The equation of a straight line on semilog paper can be written as

$$q = q_i e^{-Dt} \tag{5-1}$$

where

q = producing rate at time t, vol/unit time

q_i = producing rate at time 0, vol/unit time

D = nominal exponential decline rate, 1/time

t = time

e = base of natural logarithms, (2.718....)

Any system of units can be used as long as the product **Dt** is unitless and q and q_i are expressed in the same units. Equation 5-1 can be "derived" by stating that the decline rate at any time is proportional to the production rate, but there is no theoretical foundation for this "derivation." The theoretical foundation for exponential decline will be discussed later.

5.1.1.1 *Nominal and Effective Decline Rates*

Equation (5-1) *defines* the nominal decline rate (D). In dealing with production data, we intuitively think in terms of "effective" decline rate. For example, if we are told that a well produced 100 BOPD one year ago and now produces 50 BOPD, we naturally feel that the well declined at a rate of 50% per year. Imagine our surprise when the engineer says it is declining at 69.3% per year! Which one of the these is correct? Both of them are. Effective decline is defined as

$$D_e = \frac{q_i - q}{q_i} \tag{5-2}$$

for a given time period. The relationship between **D** and $\mathbf{D_e}$ can be derived as follows. We take **t** to be one time period (a year, perhaps). Since $\mathbf{q_i}$ and **q** are the same for both definitions of decline rate we can solve equations 5-1 and 5-2 for **q** and set the results equal:

$$q = q$$

$$q_i e^{-D} = q_i - q_i D_e$$

(t has been set to 1)

factor out q_i

$$q_i e^{-D} = q_i(1 - D_e)$$

Nominal decline as a function of effective decline is

$$D = -\ln(1 - D_e) \tag{5-3}$$

or

Effective decline as a function of nominal decline is

$$D_e = 1 - e^{-D} \qquad (5\text{-}4)$$

The authors strongly prefer the use of nominal decline rather than effective decline for reasons which will be discussed throughout the rest of the chapter.

One of the major reasons for using nominal decline has to do with changing the time units on decline rate. With nominal decline, a yearly rate can be changed to a monthly rate simply by dividing by 12. *This is not possible with effective decline!* In order to convert yearly effective rate to monthly effective rate, the *twelfth root* of 1 − D_e must be taken. Taking the twelfth root or raising a number to the twelfth power is not difficult, but it is not intuitive. An example will illustrate the above ideas.

Example 5-1
Nominal and Effective Decline Rates

Given that a well has declined from 100 BOPD to 96 BOPD during a one month period.

A) Predict the rate after 11 more months using nominal exponential decline.

B) Same as A using effective decline.

A) Using Nominal Decline

q_i = 100 BOPD

q = 96 BOPD

t = 1 month

$$D = \left[\ln\left(\frac{q_i}{q}\right)\right]/t \qquad (5\text{-}1)$$

D = .04082/mo

Find rate at end of 1 year.

$$q = q_i e^{-Dt}$$

$$q = 100e^{-.04082\,(12)}$$

q = 61.27 BOPD

B) Using Effective Decline

$$D_e = \frac{q_i - q}{q_i} \qquad (5\text{-}2)$$

$$D_e = \frac{100 - 96}{100}$$

D_e = .04/month

Convert to yearly

$$1 - D_{ey} = (1 - D_{em})^{12}$$

$$1 - D_{ey} = (1 - .04)^{12}$$

D_{ey} = .3875/year

Find rate at end of 1 year

$$q = q_i\,(1 - D_e)$$

$$q = 100(1 - .3873)$$

q = 61.27 BOPD

The authors find it much easier to use nominal decline. No matter what the units on **D** and **t**, it is only necessary to multiply by the appropriate time factor to cause the product **Dt** to be unitless. Try to predict the rate 22½ months from now using effective decline — it's not worth the effort.

5.1.1.2 *Cumulative Production*

In oil property evaluation, we are more interested in the amount of oil produced each year than the rate at any given time. In order to determine the cumulative oil production (N_p) at any time, equation 5-1 can be integrated with respect to time.

Cumulative production is the integral of rate with respect to time.

$$N_P = {}_0\int^t q\,dt$$

Substituting 5-1 for q

$$N_P = {}_0\int^t q_i\, e^{-Dt}\,dt$$

Integrating

$$N_P = \frac{q_i}{-D}\, e^{-Dt}\Big|_0^t$$

Applying the limits

$$N_P = \frac{q_i}{-D}[e^{-Dt} - e^0]$$

results in

$$N_P = \frac{q_i - q_i\, e^{-Dt}}{D}$$

recalling that $q = q_i e^{-Dt}$

$$N_P = \frac{q_i - q}{D} \qquad (5\text{-}5)$$

Again, any set of units can be used as long as time cancels out. See Example 5-2.

Example 5-2
Cumulative Production (Exponential Decline)

Calculate the amount of oil produced during the first year for the well of example 5-1.

q_i = 100 BOPD

q = 61.27 BOPD

D = .04082 (12) = .48986/year

$$N_P = \frac{q_i - q}{D} \qquad (5\text{-}5)$$

$$N_P = \frac{100 - 61.27\ \frac{STB}{day}}{.48986\ \frac{1}{yr}}$$

$$N_P = 79.06\ \frac{STB\ yr}{day}\ (365\ \frac{day}{yr})$$

$$N_P = 28{,}858\ STB$$

Some authors feel that keeping track of the units is an almost insurmountable problem. By using just a little care, however, no problems should occur.

Example 5-3 shows a technique for calculating yearly production for a number of years.

Occasionally someone will attempt to calculate production for a period by dividing the change in production rate by the *effective* decline rate. *This is wrong*. DO NOT DO IT. If you insist on using effective decline then divide the change in production rate by $-\ln(1 - D_e)$. This will result in the correct answer. Remember, the derivation of the cumulative production equation used nominal decline.

5.1.2 *Hyperbolic Decline*

Data which plots concave upward on semilog paper can often be modeled with a hyperbolic equation. The particular form of the hyperbolic equation which is most useful for decline curves is

$$q = q_i\,(1 + b\,D_i\,t)^{-1/b} \qquad (5\text{-}6)$$

where

q = producing rate at time t, vol/unit time

q_i = producing rate at time 0, vol/unit time

D_i = initial nominal decline rate (t=0), 1/time

b = hyperbolic exponent (some authors use "n")

t = time

Exponential b = Ø
Harmonic b = 1

Again, as in exponential decline, any set of units can be used as long as the product $\mathbf{D_i\ t}$ is unitless.

Socially acceptable values of b range between 0 and 1. There are, however, cases where values of b in excess of 1 can be used to fit the data. When b is equal to zero, the well is experiencing exponential decline. When b = 1, the decline is called "harmonic". Therefore, it is convenient to talk about the "b" value as though all wells were hyperbolic, with exponential and harmonic being special cases. As "b" increases, the life of the well and the remaining reserves increase dramatically.

Figure 5-1 is a plot of production rate versus time for values of b between 0 and 1. D_i was held constant at 300%/year. Notice how rapidly the curves level off at high values of b. The life of the well becomes unreasonable for large values of b. Figure 5-2 shows that a harmonically declining well will have a life 145 times the life of an exponentially declining well if both wells have the same initial decline rate. This same effect can be seen in the expected ultimate recovery, although the ratio is not as dramatic. Figure 5-3 shows that a harmonically declining well will recover 7 times as much oil as an exponentially declining well with the same initial decline rate. Notice that a value of b of .5 results in twice the reserves as a value of 0, whereas a value of 1 results in *seven* times the reserves. Since it is difficult to establish a reliable value of b during the early life of the well, values of b in excess of .5 should be adequately justified. Fortunately, the present worth of the properties does not increase with increases in b nearly as rapidly as the reserves. Figure 5-3 also contains a plot of the present worth ratio (hyperbolic versus exponential) for various values of b. The particular curve is plotted for a 10% discount rate and an initial decline rate of 300%/year. Notice that the production under harmonic decline has three times the present value of the production under exponential decline. Even though the reserves increase logarithmically, the present worth of the production increases approximately linearly.

5.1.2.1 *Nominal and Effective Decline Rates*

The quantity D_i in Equation 5-6 is the initial *nominal* decline rate at time equal 0. By definition, the decline rate changes with time so D is no longer constant as it was in exponential decline. Later in this section we will discuss how to calculate D at any time, t.

Because of the changing decline rate there are *two* possible definitions of effective decline rate in hyperbolic decline.

Example 5-3
Yearly Production Forecast (Exponential Decline)

Let's project the yearly production for the well in Example 5-1 for the next 5 years.

Step 1.

Tabulate rate at the end of each year.

The rate at the end of any year is equal to the rate at the beginning of the year times e^{-D}. Since D does not change, e^{-D} is a constant and $q_{j+1} = q_j$ times a constant for all years. The table is easily filled out. In this example,

$$q_{j+1} = q_j * e^{-.48986}$$

$$q_{j+1} = q_j * .6127$$

Step 2.

ΔN_P is equal to $q_j - q_{j+1}$ divided by D and multiplied by a time conversion factor.

Year	Rate at end of year BOPD	Yearly Production STB
0	100.00	—
1	61.27	28,858
2	37.54	17,681
3	23.00	10,834
4	14.09	6,639
5	8.64	4,061

FIGURE 5-1

Rate vs Time for Various Values of "b"

D_i - 300%/year

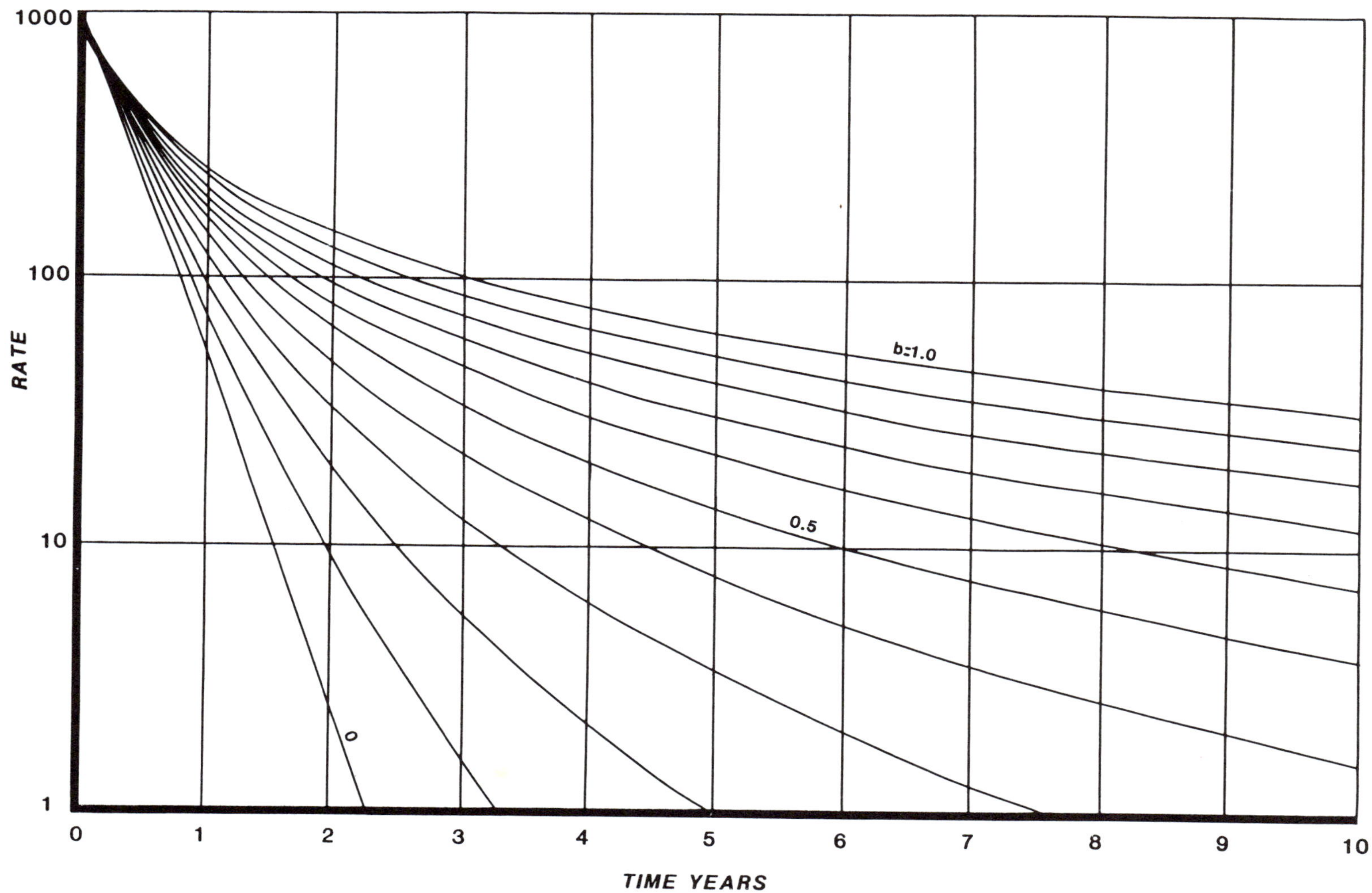

FIGURE 5-2

Ratio of life of hyperbolic well to life of exponential well with same value of D_i
$q_i / q_{el} = 1000$

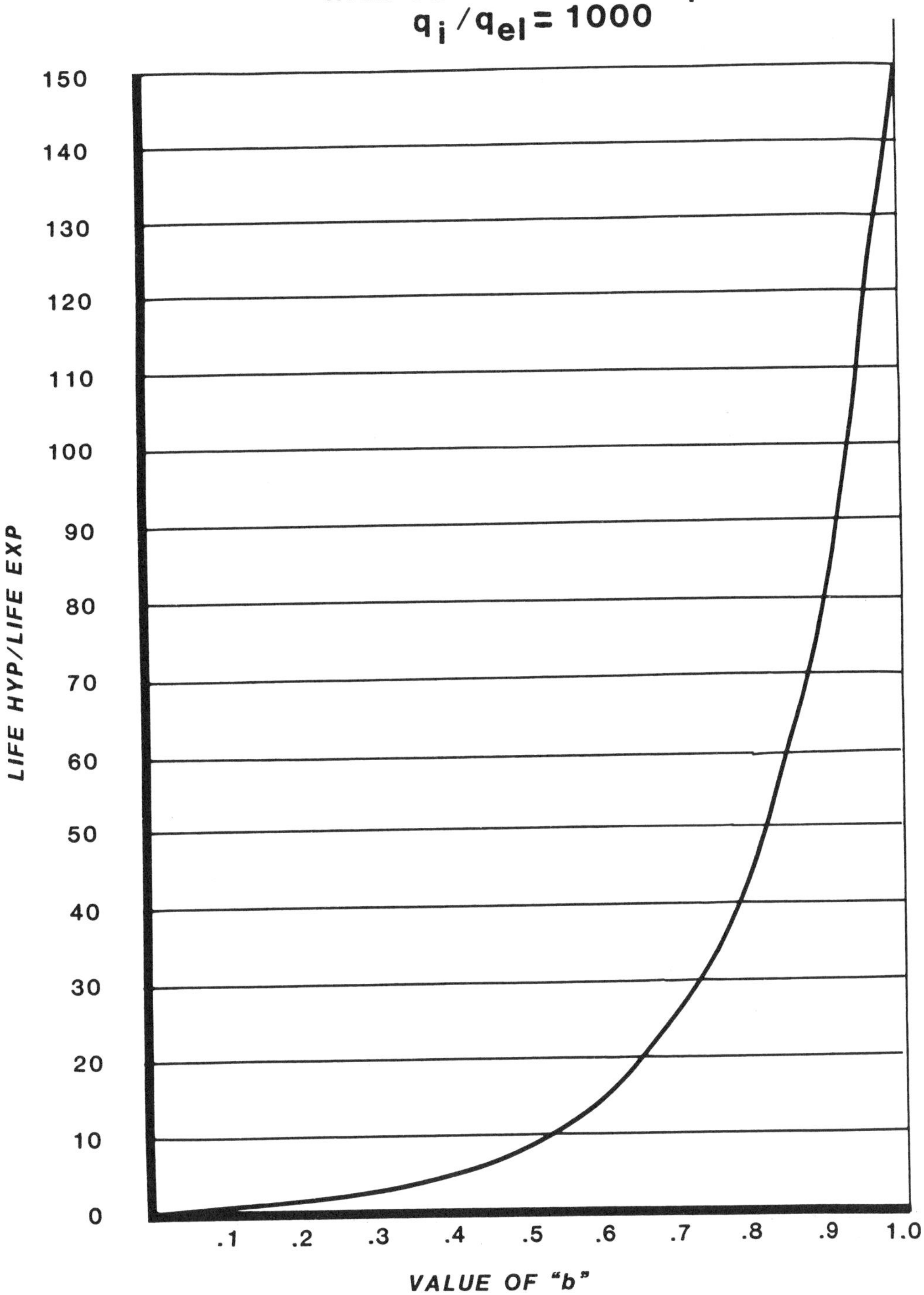

FIGURE 5-3

Ratio of Reserves Produced under Hyperbolic Decline to Reserves Produced under Exponential Decline
and
Ratio of the Present Worth of Hyperbolically Declining Well to the Present Worth of Exponentially Declining Well
(for D_i=300%/year and i=10%)
q_i/q_{el} =1000

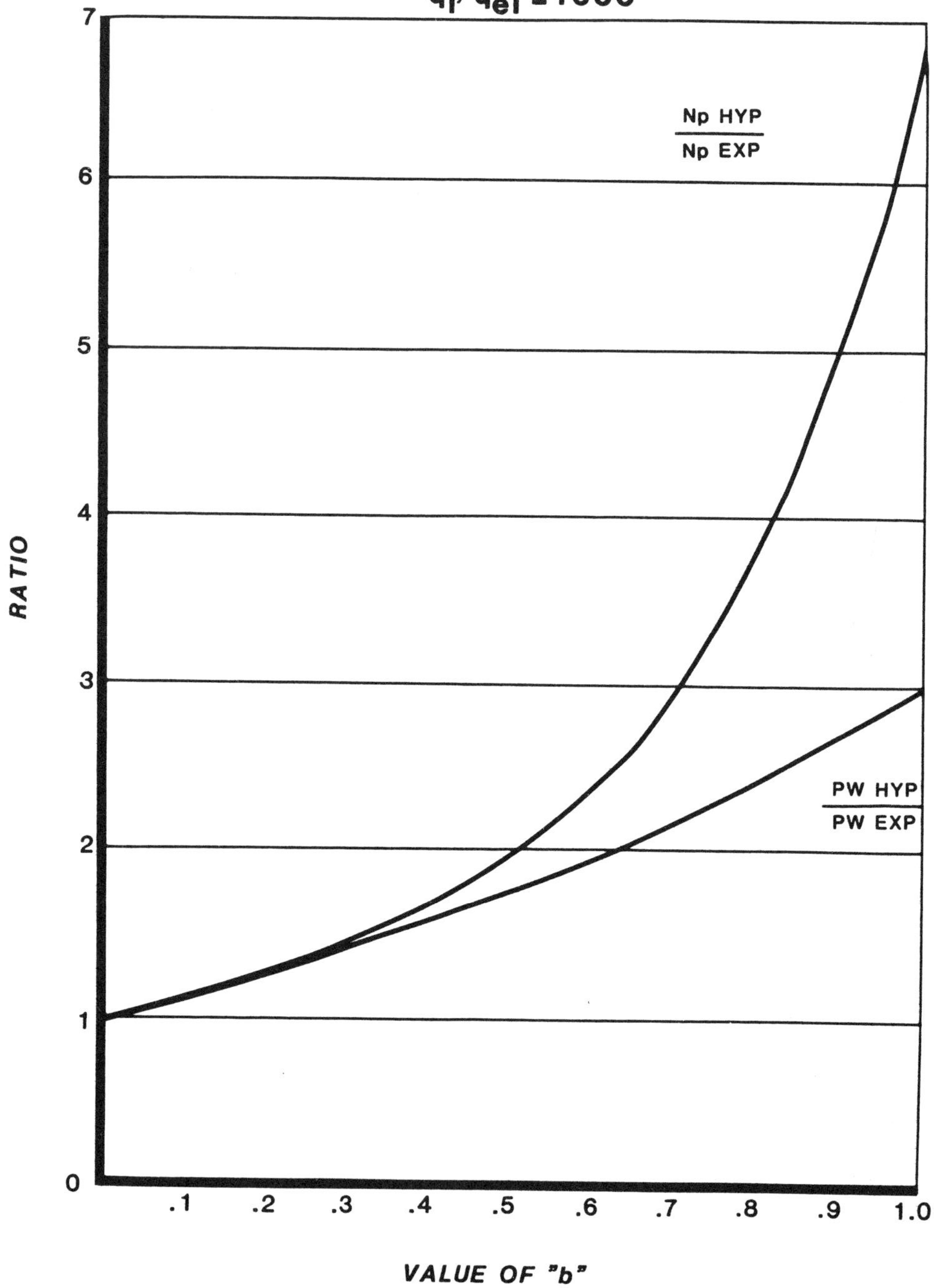

One method of defining effective decline rate is to draw a tangent to the production curve at time 0. The "effective decline rate" is calculated by the equation

$$D_{ei} = \frac{q_i - q}{q_i} \qquad (5\text{-}2)$$

where q_i and q are separated by one time period (typically one year) and are read from the *tangent to the production curve*. To calculate nominal decline rate (D_i) use Equation 5-3.

The other method of defining effective decline rate is

$$D_{ei} = \frac{q_i - q}{q_i} \qquad (5\text{-}2)$$

where q_i and q are again separated by one time period, but, in this case, both q_i and q are read from *the production curve itself*. In order to calculate initial nominal decline rate from effective decline defined in this manner, use the following equation.

$$D_i = \frac{(1 - D_{ei})^{-b} - 1}{b} \qquad (5\text{-}7)$$

These two definitions of D_{ei} can result in quite different answers for reserves. When using canned computer programs, it is often hard to determine which D should be used. Asking the salesman usually results in the answer, "Huh?". In order to find out which D has been programmed into the machine, run the following case.

Test Case
Effective date: Jan 1
Calendar year

$q_{initial}$ = 100 BOPD or 3040 BOPM

$D_{initial}$ = .5/year

b = .9

The first year reserves should be one of the following

1st year Reserves	Type of initial decline rate
29,524	Nominal
27,614	Tangent (First Method)
25,401	Actual Curve (Second Method)

The discussion in the exponential decline section about changing the units on D_i also applies to hyperbolic decline. Again, the authors find it much less confusing to work with nominal decline.

5.1.2.2 *Cumulative Production*

As with exponential decline, the cumulative production for hyperbolic decline can be determined by integrating the rate-time equation.

$$N_P = \int_0^t q\,dt$$

Substituting 5-6 for q

$$N_P = \int_0^t q_i (1 + bD_i t)^{-1/b}\,dt$$

Integrating (with $b \neq 1$)

$$N_P = \frac{1}{-\frac{1}{b} + 1} \frac{q_i}{bD_i} (1 + bD_i t)^{-\frac{1}{b} + 1} \Big|_0^t$$

Simplifying

$$N_P = \frac{b}{b - 1} \frac{q_i}{b D_i} (1 + b D_i t)^{\frac{1-b}{-b}} \Big|_0^t$$

Applying the limits

$$N_P = \frac{q_i}{(b - 1)D_i} \left[(1 + b D_i t)^{\frac{1-b}{-b}} - 1 \right]$$

Substitute $q_i^b q_i^{1-b}$ for q_i

$$N_P = \frac{q_i^b q_i^{1-b}}{(b-1)D_i}\left[(1 + b\,D_i\,t)^{\frac{1-b}{-b}} - 1\right]$$

Move q_i^{1-b} inside the brackets.

$$N_P = \frac{q_i^b}{(b-1)D_i}\left[q_i^{1-b}(1+bD_i\,t)^{\frac{1-b}{-b}} - q_i^{1-b}\right]$$

recalling that $a^x \bullet b^x = (ab)^x$
and that $a^{xy} = (a^x)^y$

$$N_P = \frac{q_i^b}{(b-1)D_i}\left[\left\{q_i(1+b\,D_i\,t)^{\frac{-1}{b}}\right\}^{1-b} - q_i^{1-b}\right]$$

since $q = q_i\,(1 + b\,D\,t)^{\frac{-1}{b}}$

$$N_P = \frac{q_i^b}{(b-1)\,D_i}\left[q^{1-b} - q_i^{1-b}\right]$$

multiplying and dividing by -1 we (finally) arrive at the equation for cumulative production

$$N_P = \frac{q_i^b}{(1-b)\,D_i}\left[q_i^{1-b} - q^{1-b}\right] \quad (5\text{-}8)$$

Equation 5-8 is valid for all values of b except b = 1. (That includes b = 0; try it.) For harmonic decline (b = 1), the equation for cumulative production is

$$N_P = \frac{q_i}{D_i}\ln\frac{q_i}{q} \quad (5\text{-}9)$$

b=1

D_i is initial nominal decline rate. If the time units on q_i and D_i are not the same, then the result must be multiplied by the appropriate factor to remove time from the answer.

Finding yearly production for a hyperbolically declining well is a little different than finding it for an exponentially declining well. Instead of finding the yearly production directly, it is easier to determine the cumulative production at the end of each year and then find the yearly production by subtracting cumulatives. Example 5-5 illustrates the principle.

Example 5-4
Cumulative Production (Hyperbolic Decline)

q_i = 100 BOPD

D_i = .5/year

b = .9

Find production for the first year

$$q = q_i(1 + b\,D_i\,t)^{-1/b} \quad (5\text{-}6)$$

$$q = 100(1 + .9 * .5 * 1)^{-1/.9}$$

q = 66.176 BOPD

$$N_P = \frac{q_i^b}{(1-b)\,D_i}\left[q_i^{1-b} - q^{1-b}\right] \quad (5\text{-}8)$$

$$N_P = \frac{100^{.9}}{(1-.9)\,.5}\left[100^{1-.9} - 66.176^{1-.9}\right]$$

$$N_p = 126.915\,[1.585 - 1.521]$$

$$N_P = 80.889\,\frac{\text{STB/day}}{1/\text{yr}}\;365\,\frac{\text{day}}{\text{yr}}$$

N_p = 29,524 STB

5.2 THEORETICAL FOUNDATION OF DECLINE CURVES

Even though decline curves had been used to forecast production since the early 1900's, it wasn't until 1956 that attention was focused on the theoretical foundation behind their use.

The following paragraphs present a brief outline of the contributions made by various authors in an effort to relate physical reservoir properties to the characteristics of the decline curves.

5.2.1 *History*

Mathews and Lefkovits investigated reservoirs producing under gravity drainage in 1956. They found that a single zone producing under gravity drainage with a free surface (secondary gas cap) had a value of b equal to 0.5. In 1958, they published another paper on gravity drainage which expanded upon their previous ideas. They found that two layers of differing height and permeability could have values of b ranging from .5 to greater than 3. The higher values of b are associated with a relatively thin, high permeability zone producing into the same wellbore as a relatively thick, low permeability zone—both under gravity drainage with a free surface. For example, a value of b of 1.43 fits a two layer zone where $h_2 = 9.26\ h_1$, and $k_1 = 97\ k_2$ for a long period of time. As the higher permeability zone depletes, the value of b approaches .5.

A well experiencing gravity drainage with no free surface (a downdip well) was shown to decline exponentially. Naturally, as up dip wells are shut in and the free surface moves down the structure, the value of b in downdip wells will again approach 0.5.

They discussed other conditions as well, but their major contribution is the demonstration of a physical system yielding a value of b of 0.5. Reading their papers is quite difficult because of their use of non-standard terminology. They refer

Example 5-5
Yearly Production Forecast (Hyperbolic Decline)

Project the yearly production for the well in Example 5-4 for 5 years.

Step 1.

Tabulate the producing rate at the end of each year using Equation 5-6

Step 2.

Calculate the cumulative production from time 0 to the end of each year using Equation 5-8 (or 5-9 if b=1).

Step 3.

Subtract the previous year's cumulative from this year's cumulative to find the production during the year.

Year	Rate of End of Year BOPD	Cumulative Production Through Year STB	Yearly Production STB
0	100.000	0	—
1	66.176	29,524	29,524
2	49.009	50,248	20,724
3	38.699	66,115	15,867
4	31.854	78,914	12,799
5	26.992	89,606	10,692

to the hyperbolic exponent as "n" which is the same symbol used by many other authors. What is radically different is their definition of n where $n = 1/b$. This is in direct conflict with most of the authors who have written about decline curves. Where Mathews and Lefkovits talk about minimum values of n being 1.45, they mean maximum values of b being 0.69, etc. (This value of "b" occurs for the sum of the production from two wells with different effective wellbore radii being plotted as one curve.) Their papers would be much more useful if they were rewritten with standard terminology.

In 1973 Fetkovich published a remarkable paper in which he showed that exponential decline is the long time solution to the diffusivity equation with constant wellbore pressure. Cox, in 1978, showed how decline curves could be used to determine various reservoir properties such as permeability, reservoir volume, and, under certain conditions, skin factor. In 1979, Ehlig-Economides and Ramey published methods for analyzing transient rate tests for wells produced at constant pressure.

These last three papers are all concerned with the same idea—the determination of reservoir properties from decline curve analysis. (By implication, if we know the reservoir properties we should be able to predict the decline characteristics.) Ehlig-Economides and Cox stayed within the bounds of theoretical reservoirs while Fetkovich branched out into empirical cases. Let's look at the theoretical reservoirs first.

5.2.2 *Assumptions Required in Derivation*

The assumptions necessary to derive the rate-time equation are as follows:

1) Horizontal, radial flow
2) Homogeneous, Isotropic Formation
3) Constant porosity
4) Pressure gradients are small
5) Single Phase Flow
6) Constant compressibility
7) Constant wellbore pressure
8) Closed exterior boundary

5.2.3 *Equations*

5.2.3.1 *Rate Time*

Under the above conditions and once pseudo-steady state flow has been established, the equation for rate vs time is:

$$q = \frac{.00708kh(P_i - P_{wf})}{\mu B\, P_{D0}} \exp\left(-\frac{.0398kt}{\phi \mu c_t A\, P_{D0}}\right) \qquad (5\text{-}10)$$

where

q = flowrate, BPD

k = permeability, md

h = thickness, ft

μ = fluid viscosity, cp

B = formation volume factor, RB/STB or RCF/SCF

ϕ = porosity, decimal

c_t = total compressibility, vol/vol/psi

$= c_o S_o + c_w S_w + c_g S_g + c_f$

A = drainage area, ft^2

t = time, days

P_{D0} = dimensionless pressure at dimensionless time = 0

$$= \frac{1}{2} \ln\left(\frac{A}{r_w^2}\right) + \frac{1}{2} \ln\left(\frac{2.2458}{C_A}\right) + S$$

r_w = wellbore radius, ft

C_A = shape factor (see pp 203-204 of "Advances in Well Test Analysis" by Earlougher)

S = skin factor

P_i = Initial reservor pressure, psi (g or a)

P_{wf} = Bottom hole flowing pressure, psi (g or a)

In terms of dimensionless variables, equation 5-10 can be expressed as:

$$q_D = \frac{1}{P_{D0}} \exp\left(\frac{-2\pi\, t_{DA}}{P_{D0}}\right) \qquad (5\text{-}11)$$

where

q_D = dimensionless producing rate

$$= \frac{141.2\, q\, \mu\, B}{k\, h\, (P_i - P_{wf})}$$

t_{DA} = dimensionless time based on drainage area

$$= \frac{.00633\, k\, t}{\phi\, \mu\, c\, A}$$

Notice that this equation is of the form

$$q = q_i\, e^{-Dt}$$

As shown by Cox, a number of formation parameters may be calculated once q_i and D are known. In order to use this method, a number of conditions have to be satisfied.

First, all of the assumptions listed in section 5.2.2 must have been met. The ones which are most likely to fail are the constant compressibility, single phase flow, and constant wellbore pressure assumptions. In order for these to be true we are limited to high pressure gas wells and oil reservoirs above the bubble point. The condition of constant wellbore pressure is most easily achieved in an oil well when the well is "pumped off". Surprisingly, exponential decline curves seem to apply below the bubble point in a number of wells. Often, we ignore the above limitations and press on.

Secondly, the value of q_i must be read from the rate-time curve by extrapolating the straight line portion backward to time 0. The pseudo steady-state portion of the curve will plot as a straight line. While the well is producing in transient flow, the rate will be greater than that predicted by the backwards extrapolation of the straight line. The transient flow period, while theoretically occurring in every well, may be hard to see in actual decline curves. For unfractured wells with reasonably high permeabilities, the transient period lasts less than a month. The transient flow period should essentially end when t_{DA} (as defined in equation 5-11) is greater than .06 for unfractured wells in the center of a circle or square. For a *highly* elongated drainage area the transient period ends when t_{DA} is greater than 2.0. For fractured wells with a highly conductive fracture, pseudo-steady state begins about $t_{DA} > 0.12$. See Earlougher's "Advances in Well Test Analysis," pp 203-204, for t_{DA} values associated with the end of transient flow. An example later in the text will calculate the time to the end of pseudo steady-state for a well.

5.2.3.2 *Original-Oil-In-Place*

Once the value of q_i and D have been found, the reservoir properties may be found by recognizing that (from equation 5-10)

$$q_i = \frac{.00708\, kh\, (P_i - P_{wf})}{\mu\, B\, P_{D0}} \qquad (5\text{-}12)$$

and

$$D = \frac{.0398\, k}{\phi \mu c_t\, A\, P_{D0}} \qquad (5\text{-}13)$$

Equations 5-12 and 5-13 may be combined to determine the reservoir volume being drained

$$Ah\phi = \frac{5.615\, q_i B_{oi}}{D\, c_t\, (P_i - P_{wf})} \qquad (5\text{-}14)$$

By knowing the original oil saturation, original oil in place (N) can be found by using

$$N = \frac{A\, h\, \phi\, S_{oi}}{5.615\, B_{oi}} \text{ (STB)} \qquad (5\text{-}15)$$

or substituting 5-14 for A h ϕ

$$N = \frac{q_i}{D\, c_t\, (P_i - P_{wf})}\, S_{oi} \qquad (5\text{-}16)$$

q_i and D must be in consistant time units.

5.2.3.3 *Permeability and Skin Factor*

Oil in place is not the only thing which can be determined from decline curves. If skin factor is known independently then permeability can be found and vice versa. By rearranging Equation 5-12.

$$k = \frac{141.2\, q_i \mu\, B\, P_{D0}}{h(P_i - P_{wf})} \qquad (5\text{-}17)$$

Since P_{D0} contains the shape factor (C_A) and the skin factor, these values must be known to estimate permeability. If an independent estimate of permeability is available (from transient testing, perhaps), then the skin factor can be determined from

$$S = \qquad (5\text{-}18)$$

$$\frac{kh}{\mu}\, \frac{(P_i - P_{wf})}{141.2\, q_i B} - \tfrac{1}{2}\ln\left(\frac{A}{r_w^2}\right) - \tfrac{1}{2}\ln\left(\frac{2.2458}{C_A}\right)$$

As Cox (1978, p. 15) states:

> "It is important to note that the determination of the skin factor [from decline curve analysis] is tied to a calculation of permeability ignoring skin, while the pore volume calculation ... is independent of skin. This is generally true of reservoir limit testing. On the other hand computation of skin effect from a transient test is tied to the porosity compressibility product, while the permeability calculation is skin independent for transient testing. Hence, when both pressure transient test data and reservoir limit test data are available, the permeability and the pore volume effective compressibility product may both be obtained independent of the skin, and the skin factor can be computed under long term conditions uninfluenced by wellbore storage, well clean-up, etc."

5.2.3.4 *Shape Factor*

Ehlig — Economides suggests that C_A can be estimated from 5-12 or 5-13 if kh/μ, A, r_w, S, and $(P_i - P_{wf})$ or ϕc_t are known. Although this is theoretically possible, it is probably asking too much to expect to be able to reliably predict the shape of the drainage area from the calculated value of C_A.

Example 5-6 illustrate the topics discussed above.

5.2.3.5 *Recovery Factor*

An expression for recovery factor can be derived from the above equations. Since ultimate recovery for an exponentially declining well is defined as

$$N_P = \frac{q_i - q_{el}}{D} \qquad (5\text{-}19)$$

and recovery factor is defined as

$$RF = \frac{N_P}{N} \qquad (5\text{-}20)$$

(N_P ← Prod oil; N ← OOIP)

substituting equations 5-19 and 5-16 into 5-20 results in

$$RF = \frac{\dfrac{q_i - q_{el}}{D}}{\dfrac{q_i\, S_{oi}}{D\, c_t\, (P_i - P_{wf})}}$$

Simplifying

$$RF = \frac{c_t\, (P_i - P_{wf})}{S_{oi}} \left(1 - \frac{q_{el}}{q_i}\right) \qquad (5\text{-}21)$$

Example 5-6
Time to pseudo-steady state

A well has the following range of estimated reservoir parameters. What is the minimum and maximum time to pseudosteady state?

$$10 < k < 50 \text{ md}$$
$$.12 < \phi < .16$$
$$.5 < \mu < .7$$
$$.6 < S_o < .7$$
$$15\text{x}10^{-6} < C_o < 20\text{x}10^{-6} \text{ psi}^{-1}$$
$$.3 < S_w < .4$$
$$3\text{x}10^{-6} < C_w < 6\text{x}10^{-6} \text{ psi}^{-1}$$
$$4\text{x}10^{-6} < C_f < 15\text{x}10^{-6} \text{ psi}^{-6}$$
$$40 < A < 80 \text{ acres}$$

Assuming the well is in the center of a regular shaped drainage area, $t_{DA} > .06$ will suffice.

$$C_{t\ min} = C_o S_o + C_w S_w + C_f$$
$$= .6 * 15\text{x}10^{-6} + .4 * 3\text{x}10^{-6} + 4 \text{ x } 10^{-6}$$
$$C_{t\ min} = 14.20 \text{ x } 10^{-6} \text{ psi}^{-1}$$
$$C_{t\ max} = .7 * 20\text{x}10^{-6} + .3 * 6 \text{ x}10^{-6} + 15\text{x}10^{-6}$$
$$= 30.8 \text{ x } 10^{-6} \text{ psi}^{-1}$$
$$t_{min} = \frac{.06\ \phi\ \mu\ C_t\ A}{.00634\ k}$$
$$= \frac{(.06)(.12)(.5)(14.2\text{x}10^{-6})(40\text{x}43560)}{(.00634)(50)}$$
$$t_{min} = .28 \text{ days}$$
$$t_{max} = \frac{(.06)(.16)(.7)(30.8\text{x}10^{-6})(80\text{x}43560)}{(.00634)(10)}$$
$$t_{max} = 11.4 \text{ days}$$

If the well is indeed produced at a constant bottom hole pressure, pseudo-steady state will occur in less than a month.

This recovery factor should come as no shock—it is the recovery factor for a reservoir with an initial pressure of P_i and a final reservoir pressure of P_{wf} producing constant compressibility fluids (i.e. above the bubble point). The recovery factor for oil above the bubble point will calculate to be less than 10% of original oil-in-place, which is to be expected.

5.2.3.6 *Decline Rate*

If q_{el} is assumed to be zero, D can be related to the initial rate and the estimated ultimate recovery.

Rearranging Equation 5-16

$$D = \frac{q_i}{N \frac{c_t}{S_{oi}} (P_i - P_{wf})} \qquad (5\text{-}22)$$

assuming $q_{el} = 0$, substitute 5-21

$$D = \frac{q_i}{(N)(RF)} \qquad (5\text{-}23)$$

or

$$D = \frac{q_i}{N_P} \qquad (5\text{-}24)$$

This result has also been derived by Fetkovich.

So much for methods based on strict theory. They should work well above the bubble point and could be very useful in estimating original oil in place. When combined with properly run transient tests, a great deal of information about the reservoir can be derived. It should be remembered that all of the above discussion depends upon the well producing with a constant (and known) flowing bottom hole pressure. If this condition is not met, then it may be possible to use superposition to derive answers. The key will be knowing flowing bottom hole pressure as a function of time.

5.3 DECLINE CURVE ANALYSIS

5.3.1 *General*

5.3.1.1 *Plotting the Data*

Normally, producing rate is plotted every month on semilog graph paper. The scale for production can be Bbls per month or Bbls per day (either calendar or producing day). No matter what scale is used, the rate should be plotted as a point in the *middle* of the month. There is both a practical and a theoretical reason for doing this. From a practical standpoint, if a data point is plotted on the line between December and January, it is impossible to tell which month it represents. By plotting in the middle of the month, this problem goes away. From a theoretical standpoint, we are plotting one point which represents the average production rate for the period. We would like that point to fall on the smooth curve through the data which represents the *instantaneous* production rate at any time. For exponential decline the point in time when the average rate and the instantaneous rate are equal (t_x) can be derived as follows.

For units on q of bbls/year and D of 1/year we want to calculate the time (t_x) when:

$$\Delta N_P = q$$

$$\frac{q_i(1 - e^{-D})}{D} = q_i \, e^{-Dt_x}$$

q_i cancels

$$e^{-Dt_x} = \frac{(1 - e^{-D})}{D}$$

FIGURE 5-4

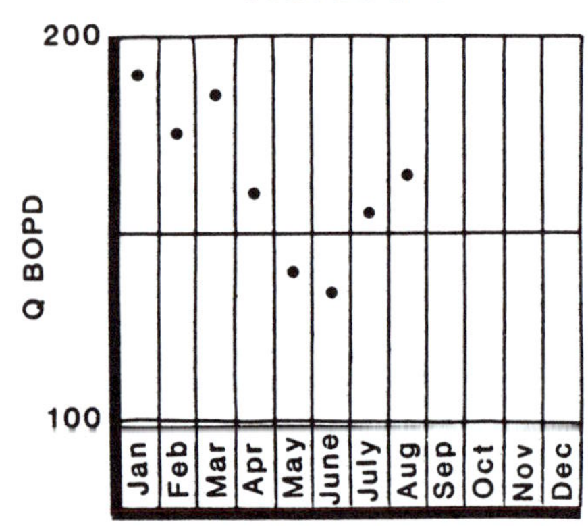

$$t_x = \frac{\ln\left(\frac{1 - e^{-D}}{D}\right)}{-D} \qquad (5\text{-}25)$$

An expression can be derived in a similar manner for hyperbolic decline. Table 5-1 contains the values of t_x for various values of D_i and b. Notice that even for extreme values of D_i and b, the average production rate should be plotted near the center of the period. This is true whether production is being plotted monthly or yearly.

For an oil well, the gas and water data are often plotted as gas-oil ratio and water-oil ratio, although sometimes they are plotted as rates (MCFPD or BWPD). For a gas well, condensate data may be plotted as liquid-gas ratio (bbls/MMSCF) or as a production rate (BCPD). If the well is one you have never seen before, it is wise to plot the data as many ways as possible. This will often highlight ''boiler-housed'' data. A perfectly constant GOR for six months is often a sign that the data has been fabricated. On wells where gas is not being sold, using a constant GOR is quite common. Water data is often very suspect. Since water usually isn't sold, no one likes to spend the time and money necessary to get good data. Unfortunately good water data is often required to predict performance in water drive reservoirs. If the water is being hauled for disposal, it is sometimes possible to get good data from lease cost records.

5.3.1.2 *Correcting for Downtime*

Often a well is down (doesn't produce) for part of a month. If production rate in barrels per month is being plotted, the data point will be quite low. One way to remove the effect of downtime is to plot barrels per producing day. This usually smooths out the curve quite nicely. Of course, unless someone records downtime you can't correct for it. Also, when making the production forecast from the decline curve plotted on a per producing day basis, remember that you are assuming zero downtime. The calculated yearly production should be reduced by the anticipated future downtime.

5.3.1.3 *Multiple Wells on a Lease*

It is not unusual for production to be recorded on a lease basis rather than by well. When this occurs it is sometimes helpful to plot BOPM per well. This is a pretty shakey way to use decline curves, but at this point you're probably pretty desperate. Try it, it may be better than the alternatives.

Table 5-1
Time during period when average production rate equals instantaneous production rate.

D_i (1/yr)	b					
	0	.2	.4	.6	.8	1.0
0.00	.5000	.5000	.5000	.5000	.5000	.5000
0.25	.4896	.4878	.4861	.4845	.4829	.4823
0.50	.4792	.4762	.4734	.4709	.4685	.4663
0.75	.4689	.4652	.4618	.4588	.4561	.4536
1.00	.4587	.4545	.4511	.4480	.4452	.4427
1.25	.4486	.4446	.4412	.4382	.4355	.4332
1.50	.4386	.4350	.4319	.4292	.4268	.4247
1.75	.4289	.4259	.4232	.4210	.4189	.4171
2.00	.4193	.4171	.4151	.4133	.4117	.4102

5.3.1.4 *Wild Fluctuations*

Sometimes sales figures are reported rather than production figures. On leases which are being trucked and are producing about 1-1/2 truckloads per month, the "production" is often reported in truckloads, as 2, 1, 2, 1, 2, 1, etc. This leads to a rather erratic decline curve. Fluctuations can sometimes be smoothed out by plotting a 3-month running average. In that case the production plotted in each month (j) is:

$$q_j = \frac{q_{j-1} + q_j + q_{j+1}}{3} \quad (5\text{-}26)$$

Five month, and seven month, etc. running averages can also be plotted, but they are not recommended. If you plot all the data as a single point it's hard to find an unique answer.

If more than one year of production data is available, it is often helpful to plot the data on a yearly basis rather than monthly. If this is done, the data must be plotted near the middle of the year. Refer to the discussion on plotting the data made earlier in this chapter.

5.3.2 *Exponential Decline*

The objective of analyzing decline curves is to attempt to predict the future production from a well. In order to do this one must obtain q_i, D, and b as of the effective date of the study. For exponential decline, the process is quite easy. For hyperbolic decline, it can be somewhat difficult.

5.3.2.1 *Rate-Time Plots*

With exponential decline, after the data is plotted on a piece of semilog paper, a straightedge is used to draw the best "eyeballed" curve through the data. It is recommended that this be done by hand rather than by a least squares technique because spurious data points can be ignored and judgement can be applied. Once the line is drawn, two points should be picked near the extremities of the line. Call them (t_1, q_1) and (t_2, q_2). The decline rate is then found by solving the equation:

$$D = \frac{\ln\left(\frac{q_1}{q_2}\right)}{t_2 - t_1} \quad (5\text{-}27)$$

Note that the units on D are the same as the units on t. Since we are dealing with nominal decline, the units on D can be changed by multiplying by the appropriate time conversion factor (12 mos/yr, etc).

To find q at the effective date, read it from the graph. Or, if you wish to calculate it, use the point (t_1, q_1) from above and calculate q from

$$q_{effdat} = q_1 \, e^{-D(t_{effdat} - t_1)} \quad (5\text{-}28)$$

Watch the units on D and t!

5.3.2.2 *Rate-Cum Plots*

By rearranging equation 5-5 it can be seen that a plot of producing rate versus cumulative production will be a straight line on Cartesian coordinate paper **for exponential decline.** This can be derived as follows:

$$N_P = \frac{q_i - q}{D} \quad (5\text{-}5)$$

$$q = q_i - D\,N_P \quad (5\text{-}5a)$$

Equation 5-5a is of the form y = b + mx where:

b (the intercept) = q_i

m (the slope) = $-D$

A plot of rate versus cum will look like

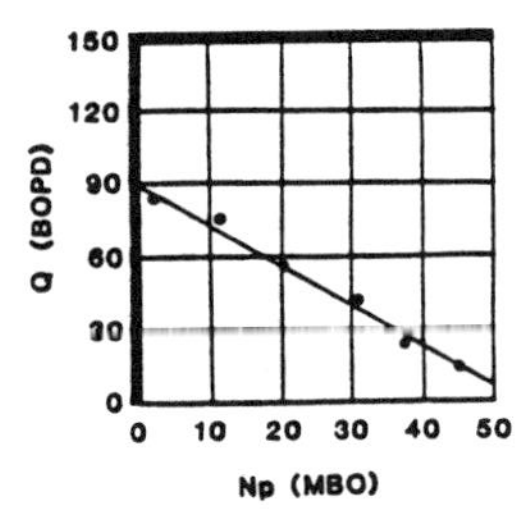

The decline rate is calculated from rise over run. It will have the same time units as the time units on q (1/day for BPD, 1/mo. for BPM, etc.). Since this D is **nominal** exponential decline, simple multiplication can be used to convert to any desired time units (D 1/day * 365 = D 1/yr). Theoretically, this decline rate should be exactly the same as the one calculated from a rate-time plot. In practice it seldom is because of a difference in ''eyeballing'' the curves.

This plot is quite handy if you wish to evaluate a frac job or an acid job. If you are willing to live with the assumption that the ultimate recovery is the same, then a frac job merely increases the current rate. The plot will appear as follows:

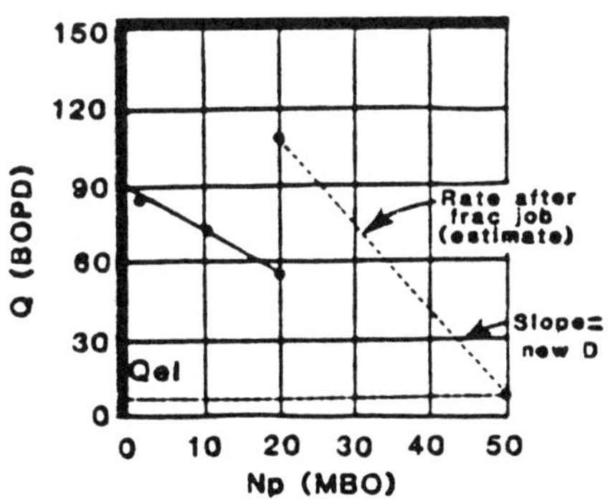

You will have to determine the post frac rate independently, but after that the plot will give you the new decline rate.

5.3.3 *Hyperbolic Decline*

There are a number of methods to solve for q_i, D_i, and b when hyperbolic decline is assumed (concave upward on semilog plot). Some of the more useful ones are:

1. Type Curve Analysis

2. Trial and Error

3. Find D_i from a tangent

4. Special graph paper

Other techniques have been used over the years, but only those shown above will be discussed.

5.3.3.1 *Fetkovich Type Curves*

Fetkovich combined the early transient flow period with the empirical hyperbolic equations. The resulting curves are plotted on log-log paper and can be used to determine q_i, D_i, and b if the well is producing under pseudo-steady state con-

Example 5-7
Accounting for Downtime
Plotting the Data

Given the following production data:

Month	Production (Bbls)	Cumulative (Bbl)	Calendar Days	Bbl Per Calendar Day	Days Prod.	Bbl Per Prod. Day
3/82	5607	5607	31	180.9	23	243.8
4/82	6261	11868	30	208.7	27	231.9
5/82	5736	17604	31	185.0	26	220.6
6/82	6084	23688	30	202.8	29	209.8
7/82	5190	28878	31	167.4	26	199.6
8/82	5697	34575	31	183.8	30	189.9
9/82	5057	39632	30	168.6	28	180.6
10/82	5154	44786	31	166.3	30	171.8
11/82	4085	48871	30	136.2	25	163.4
12/82	4665	53536	31	150.5	30	155.5
			306		274	

Example 5-7
(Continued)

A. Plot barrels of oil per month (BOPM) versus time on semilog paper. Draw your best fit for the actual production. Assume exponential decline. Calculate the amount of production from 3/1/82 to 12/31/82 using your best fit and compare to the actual production.

B. Plot barrels of oil per calendar day versus time on semilog paper. Draw your best fit for the actual production. Assume exponential decline. Calculate the amount of production from 3/1/82 to 12/31/82 using your best fit and compare to the actual production.

C. Plot barrels of oil per producing day versus time on semilog paper. Draw your best fit for the actual production. Assume exponential decline. Calculate the amount of production from 3/1/82 to 12/31/82 using your best fit and compare to the actual production for this period.

1. Ignore downtime.
2. Incorporate downtime.

D. Which approach for plotting the data is correct?

Solution:

Note on Figure 5-5, part A, B, and C, that all production is plotted at the midpoint of the month since we are plotting average monthly data. The rates in the decline curve equations are instantaneous rates. When data is plotted at the midpoint of the month, we are assuming that the average monthly rate is a good approximation of the instantaneous rate at the midpoint of the month. This is a good assumption, especially at low decline rates and low values of "b" (refer to Table 5-1).

A. Refer to plot, Example 5-7, Part A.

Using the best fit shown:

$Q_{3/1/82} = 6200$ BOPM

$Q_{3/1/85} = 1800$ BOPM

$t = 3$ years

$$D = \frac{\ln\left(\frac{6200}{1800}\right)}{3}$$

$D = 0.412$/year

1. Calculate $Q_{12/31/82}$

$Q_{12/31/82} = 6200\ e^{-(0.412 * 10/12)}$

$Q_{12/31/82} = 4398.3$ BOPM

Example 5-7
(Continued)

2. Calculate production for 3/1/82 — 12/31/82 period using best fit.

$$Np = \frac{(Q_{3/1/82} - Q_{12/31/82})}{0.412} \; 12$$

$$= 52477 \text{ bbl.}$$

3. Compare to actual production.

Actual = 53536 bbl.
Estimated = 52477 bbl.

This calculation is performed to demonstrate how well our best fit would have predicted the actual production. In cases where the EUR is calculated, always use actual data when possible. (Remember EUR = CUM + ERR where CUM is actual production.)

B. Refer to plot, Example 5-7, Part B.

Using the best fit shown:

$Q_{3/1/82}$ = 200 Bbl./Calendar day

$Q_{3/1/85}$ = 62 Bbl./Calendar day

t = 3 years

$$D = \frac{\ln\left(\frac{200}{62}\right)}{3}$$

$$D = 0.390/\text{year}$$

(This value for D would be exactly the same as the D from Part A if our calibrated eyeball was consistent.)

1. Calculate $Q_{12/31/82}$

$$Q_{12/31/82} = 200\ e^{-(0.390 * 10/12)}$$

$$= 144.5 \text{ BOPD}$$

2. Calculate production for 3/1/82 — 12/31/82 period using best fit.

$$Np = \frac{(Q_{3/1/82} - Q_{12/31/82})}{0.390} \; 365$$

Example 5-7
(Continued)

$$Np = \frac{(200 - 144.5)\ 365}{0.390}$$

$$= 51942$$

3. Compare to actual production.

   Actual = 53536 bbl.
   Estimated = 51942 bbl.

C. Refer to plot, Example 5-7, Part C.

Using best fit shown:

$Q_{3/1/82} = 250$ BOPD

$Q_{3/1/85} = 40$ BOPD

$t = 3$ years

$$D = \frac{\ln\left(\frac{250}{40}\right)}{3}$$

$D = 0.611$/year

1. Calculate $Q_{12/31/82}$

   $Q_{12/31/82} = 250\ e^{-0.611 * 10/12}$

   $= 150.2$ BOPD

2. Calculate production for the period 3/1/82 to 12/31/82 using the best fit.

   a. Ignore downtime.

   $$Np = \frac{(250 - 150.2)\ 365}{0.611}$$

   $Np = 59619$

Example 5-7
(Continued)

b. Incorporate downtime.

$$\text{Downtime factor} = \frac{\text{Days produced}}{\text{Days available}}$$

$$= \frac{274}{306}$$

$$= 0.894$$

$$Np = \frac{(250 - 150.2)}{0.611/\text{year}} \times 365 \frac{\text{day}}{\text{year}} \times 0.894 \frac{\text{days produced}}{\text{days available}}$$

$Np = 53{,}299$ bbls.

3. Compare to actual.

a. Ignore downtime.

When production is plotted on a per producing day, the data is smoothed and it is often easier to determine the best fit. However, using the best fit will automatically overstate the results if there is downtime.

Actual = 53536 bbl.
Estimated = 59619 bbl.

b. Incorporate downtime.

This results in the best comparison of actual versus estimated and hopefully would give the best estimate for future production.

Actual = 53,536
Estimated = 53,299

D. Parts A, B, and C(2b) will all give the same results if in Parts A and B your best fit is an accurate weighting of the data points. When production is erratic because of downtime, Part C(2b) will likely give better results since it will be easier to fit the data. However, downtime must be taken into account when projecting production from data plotted on a barrels per producing day basis.

Example 5-7, Part A

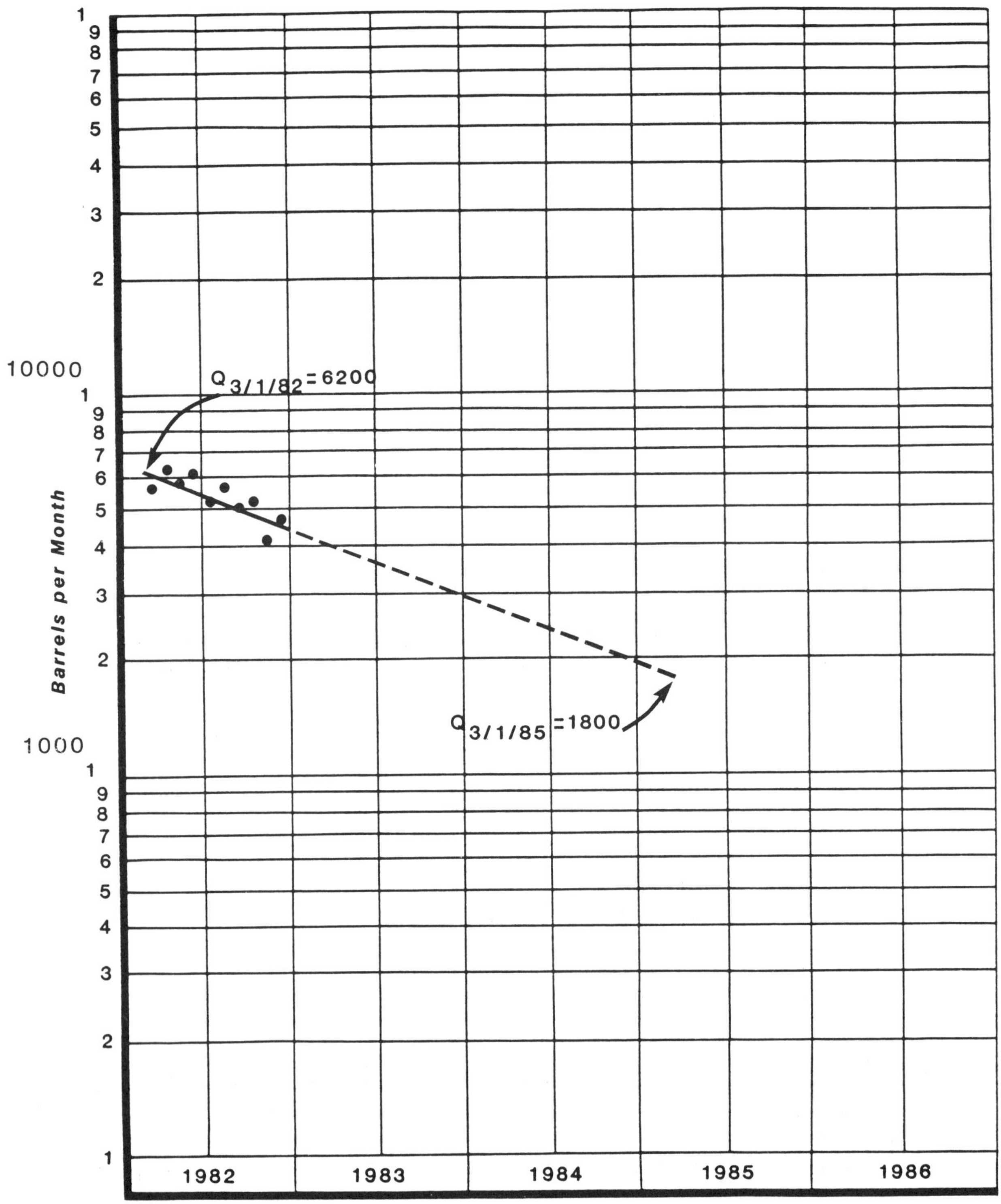

Example 5-7, Part B

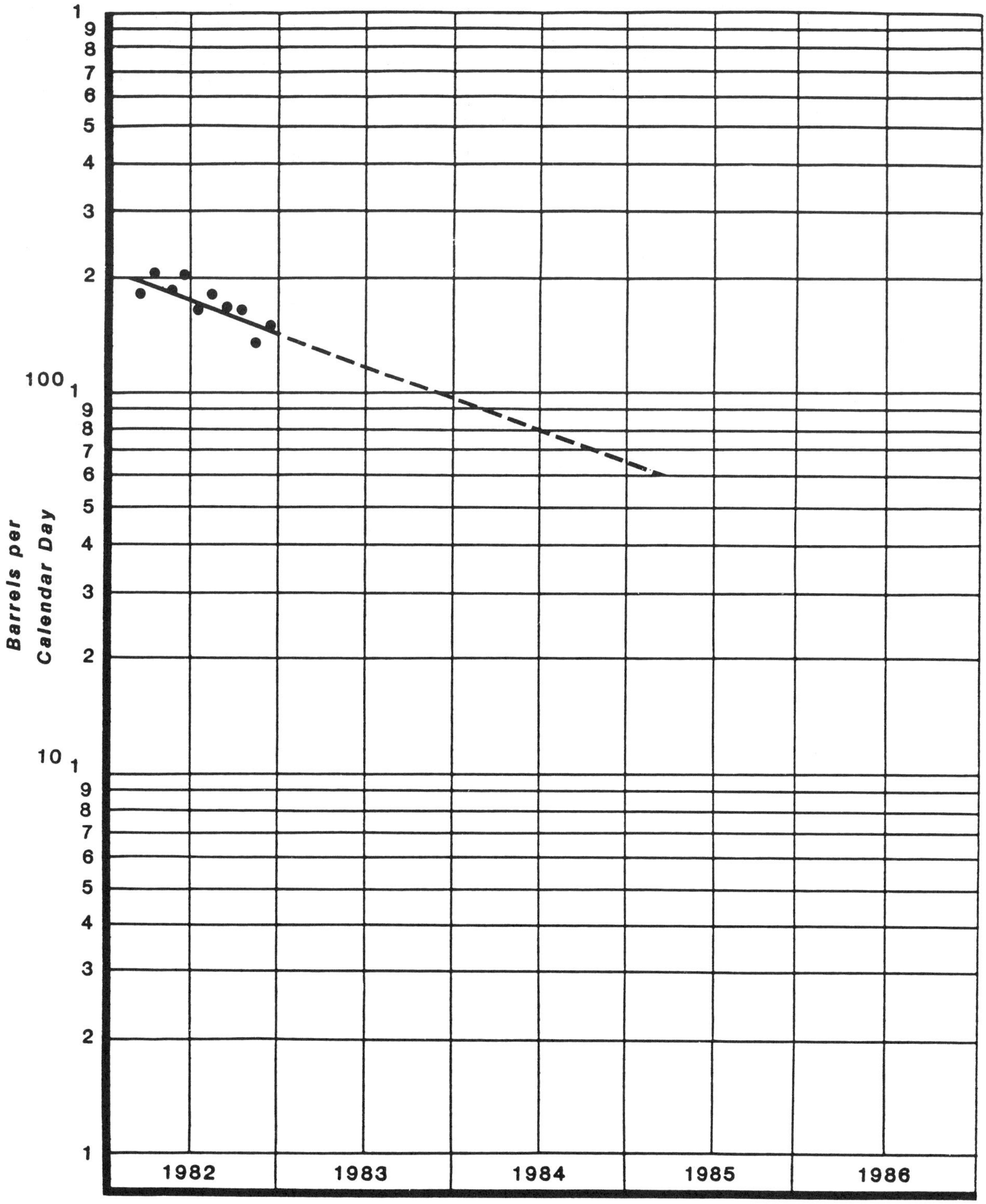

Example 5-7, Part C

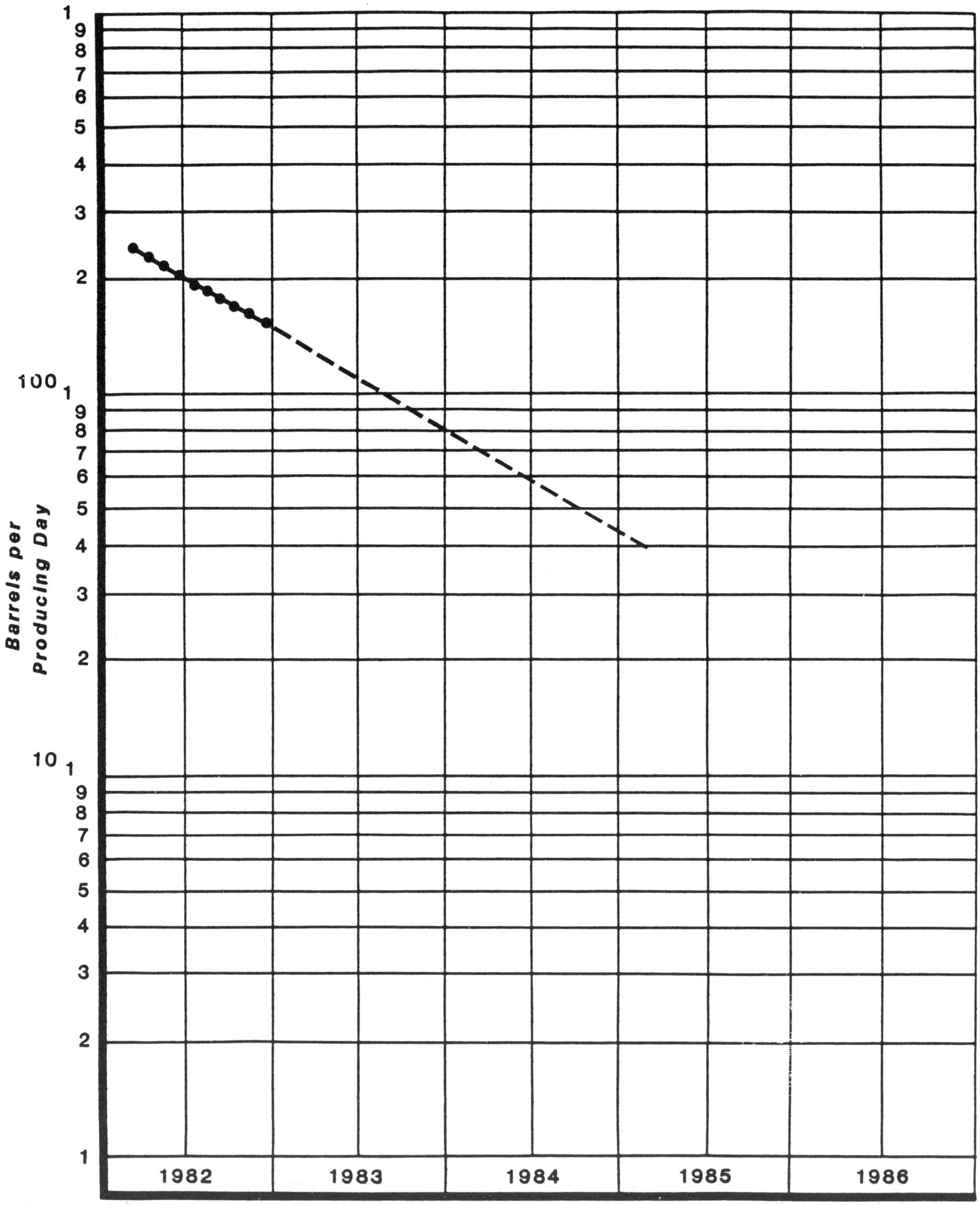

ditions. If the well is in transient flow the curves can be used to determine kh/μ, ϕc_t, and r_e/r_w. Fetkovich's curves are valid only for circular, bounded reservoirs with a well in the center. Curves for other geometries could be plotted. Fetkovich's type curve can be derived from the equation for the pressure behavior of a well in a closed reservoir producing at pseudo-steady state.

In the back of this book is a copy of Fetkovich's type curve. The dimensionless parameters are defined by

$$t_{Dd} = \frac{t_D}{\frac{1}{2}\left[\left(\frac{r_e}{r_w}\right)^2 - 1\right]\left[\ln\left(\frac{r_e}{r_w}\right) - \frac{1}{2}\right]} \tag{5-29}$$

$$q_{Dd} = q_D \left[\ln\left(\frac{r_e}{r_w}\right) - \frac{1}{2}\right] \tag{5-30}$$

To use type curve matching, overlay the type curve with a sheet of tracing paper, trace the major grid lines, and label the axes of the tracing paper with time and rate scales appropriate for the data you are going to plot. Usually the time scale will be in months from 0.1 month and the rate scale will be in BOPM from 10 or 100, or in BOPD from 1. With high rate wells, naturally the rate scale will begin higher. Now, using the type curve to provide intermediate lines, plot rate versus time. Type curves are just like semilog curves—plot the data in the middle of the month. i.e. at .5, 1.5, 2.5 months, etc. Do not connect the dots. Ignore the curves and scale on the type curve during plotting; use only the grid on the tracing paper. Once all of the data has been plotted, and *keeping the grids parallel*, slide the tracing paper around on the type curve until the data points match one of the type curves. Usually more "weight" is given to later data points.

Once a match has been made, trace the matched curve and pick a "match point". The match point can be anywhere on paper; it doesn't have to be on the curve. Usually, an intersection of the major grid lines on the tracing paper is used for convenience. Record the values from the tracing paper (q and t) and the corresponding values lying beneath that point on the type curve grid (q_{Dd} and t_{Dd}).

Earlougher's book has a good section on type curve matching (p.24-25).

Depending upon whether the data points matched in the analytical or empirical region, use one of the following procedures to determine the appropriate parameters.

5.3.3.1.1 *Analytical Region—find kh/μ,* r_e/r_w, ϕc_t

r_e/r_w is obtained from the curve which was matched. Convert q to BOPD. Then kh/μ is calculated from q and q_{Dd} at the match point using the equation

Example 5-8
Hyperbolic Decline — Type Curve

The production data, semilog plot, and log-log plot for a hyperbolic well (presented at the end of this example) are designed to demonstrate the mechanics of type curve analysis for decline curves. The type curve is included in the back pocket. Plot the data (pages 5-31 and 5-32) and verify the following match points. Your plot should look like the log-log plot on page 5-30.

Match Point No. 1	Match Point No. 2
t = 1 mo.	t = 10 mo.
q(t) = 10,000	q(t) = 1000
t_{Dd} = .042	t_{Dd} = 0.42
q_{Dd} = 3.3	q_{Dd} = 0.33

Example 5-8
(Continued)

To solve for b, q_i, and D_i:

1. Place transparent log-log overlay on top of the type curve so that our match point values of q_D and t_{Dd} on the type curve correspond with the match point values of q and t for the actual plotted values of production.

   Although only one match is required, two match points have been given. This will insure that the horizontal and vertical axes of the log-log plot and type curve are parallel (maybe).

2. Once the match has been made, notice that the production on the log-log plot corresponds with the type curve having a "b" value of 0.6. This is where the application of science ends and art begins. You probably could find a match for other values of "b". This is true even with 4 years of production history. It is almost impossible to come up with a good match with less than 3 years of production history unless the initial decline rate is quite high.

3. Now, calculate the values of q_i and D_i.

   a. For the first match point.

   $$q_{Dd} = \frac{q(t)}{qi}$$

   $$t_{Dd} = D_i t$$

   $q(t) = 10{,}000$ BOPM $\qquad q_{Dd} = 3.3$

   $t = 1$ mo. $\qquad t_{Dd} = 0.042$

   $$qi = \frac{10{,}000}{3.3} = 3030 \text{ BOPM}$$

   $$D_i = \frac{0.042}{1} = 0.042/\text{mo.}$$

   b. For the second match point.

   $q(t) = 1000$ BOPM $\qquad q_{Dd} = 0.33$

   $t = 10$ mo. $\qquad t_{Dd} = 0.42$

   $$qi = \frac{1000}{0.33} = 3030 \text{ BOPM}$$

   $$D_i = \frac{0.42}{10} = 0.042/\text{mo.}$$

   Notice that both match points gave the same result. If the time t is increased by a factor of 10, dimensionless time t_{Dd} is also increased by a factor of 10, thus the value of D_i is the same. The same concept is true for different values of q(t) and q_{Dt}.

   Thus, once a good "fit" is found for the log-log plot of the production data and one of the curves on the type curve, the match point can be taken anywhere on the plot that is convenient.

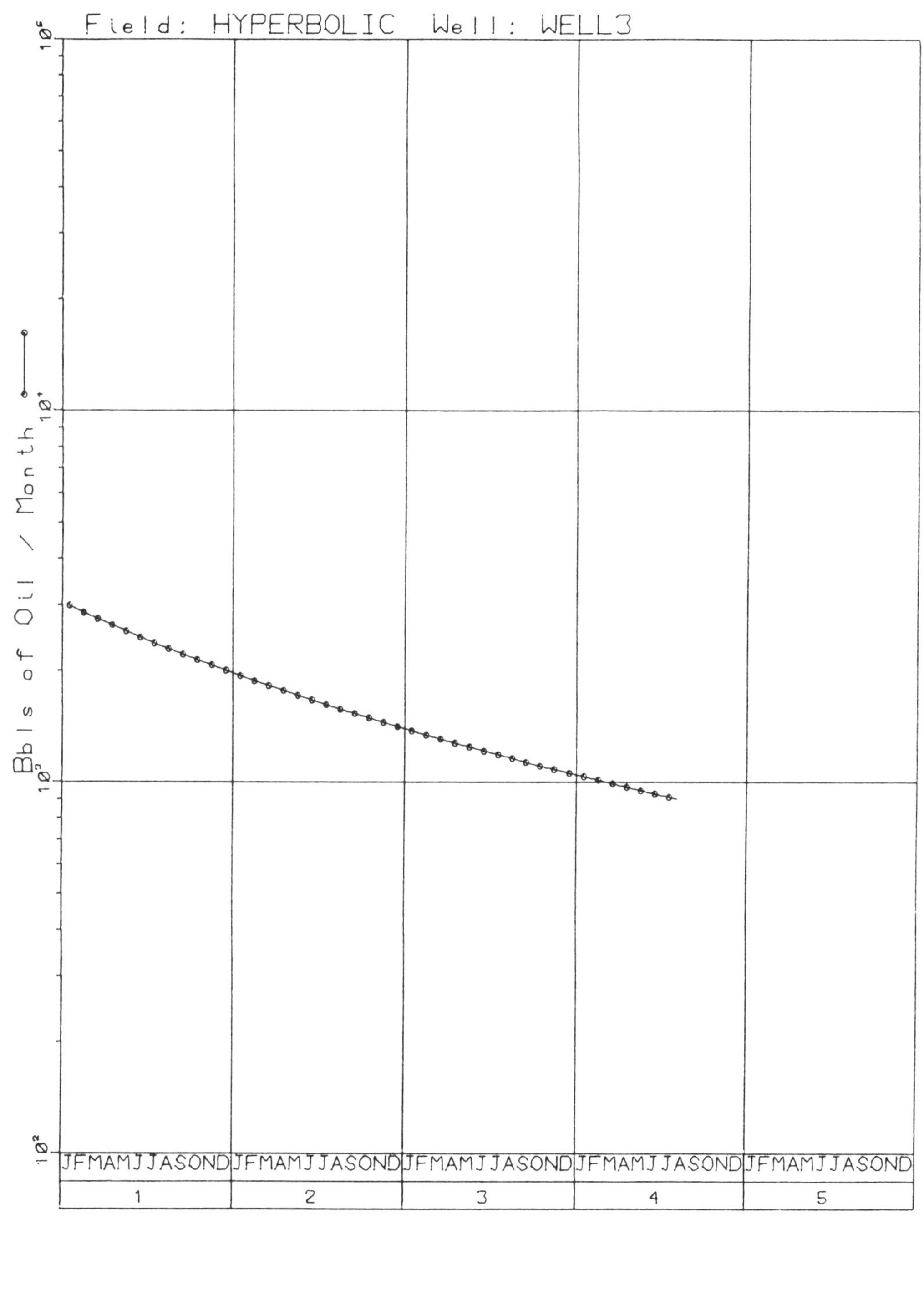
Field: HYPERBOLIC Well: WELL3
Bbls of Oil / Month
10⁵
10⁴
10³
10²
JFMAMJJASOND JFMAMJJASOND JFMAMJJASOND JFMAMJJASOND JFMAMJJASOND
1 2 3 4 5

Log-Log Plot
Hyperbolic Well 3
1/1 Through 12/4 (mo/year)

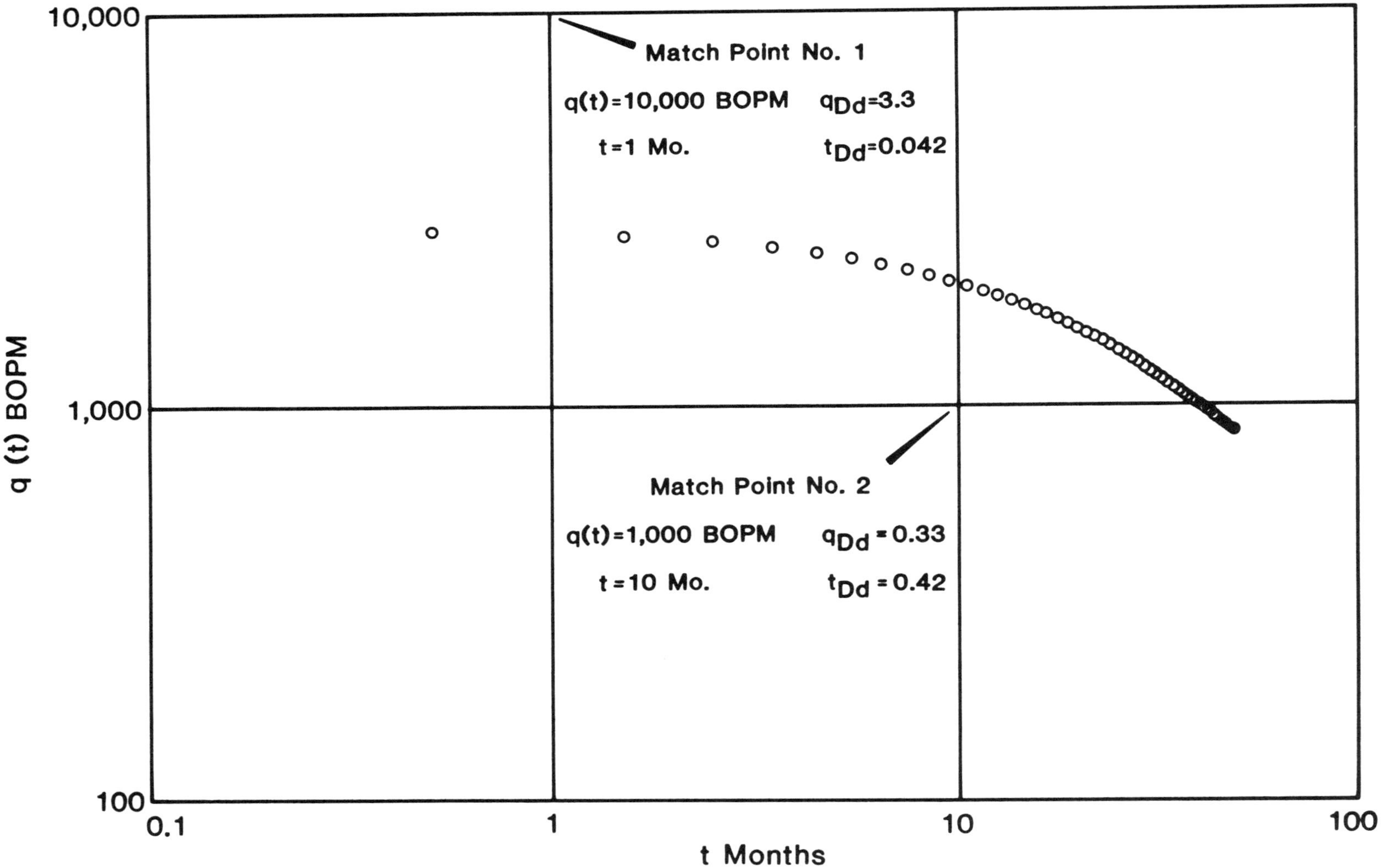

Production Data for Example 5-8
HYPERBOLIC WELL3

				Oil/Cond Bbls	
Mo	Year	Status	Day*	Month	Cum
1	1	OPR	96.1	2978	2978
2	1	OPR	102.1	2859	5837
3	1	OPR	88.6	2748	8585
4	1	OPR	88.1	2644	11229
5	1	OPR	82.1	2545	13774
6	1	OPR	81.8	2453	16227
7	1	OPR	76.3	2365	18592
8	1	OPR	73.6	2283	20875
9	1	OPR	73.5	2205	23080
10	1	OPR	68.7	2131	25211
11	1	OPR	68.7	2062	27273
12	1	OPR	64.4	1995	29268
ST	1			29268	
1	2	OPR	62.3	1932	31200
2	2	OPR	66.9	1872	33072
3	2	OPR	58.6	1816	34888
4	2	OPR	58.7	1761	36649
5	2	OPR	55.2	1710	38359
6	2	OPR	55.3	1660	40019
7	2	OPR	52.0	1613	41632
8	2	OPR	50.6	1568	43200
9	2	OPR	50.8	1525	44725
10	2	OPR	47.9	1484	46209
11	2	OPR	48.2	1445	47654
12	2	OPR	45.4	1407	49061
ST	2			19793	
1	3	OPR	44.2	1371	50432
2	3	OPR	47.7	1336	51768
3	3	OPR	42.0	1303	53071
4	3	OPR	42.4	1271	54342
5	3	OPR	40.0	1240	55582
6	3	OPR	40.4	1211	56793
7	3	OPR	38.1	1182	57975
8	3	OPR	37.3	1155	59130
9	3	OPR	37.6	1128	60258
10	3	OPR	35.6	1103	61361
11	3	OPR	35.9	1078	62439
12	3	OPR	34.0	1055	63494
ST	3			14433	

Production Data for Example 5-8
HYPERBOLIC WELL3

Mo	Year	Status	Day*	Oil/Cond Bbls Month	Oil/Cond Bbls Cum
1	4	OPR	33.3	1032	64526
2	4	OPR	34.8	1010	65536
3	4	OPR	31.9	988	66524
4	4	OPR	32.3	968	67492
5	4	OPR	30.6	948	68440
6	4	OPR	30.9	928	69368
7	4	OPR	29.4	910	70278
8	4	OPR	28.8	892	71170
9	4	OPR	29.1	874	72044
10	4	OPR	27.6	857	72901
11	4	OPR	28.0	841	73742
12	4	OPR	26.6	825	74567
ST	4			11073	

*Per Calendar Day

$$\frac{kh}{\mu} = \frac{q}{q_{Dd}} \frac{141.2\,B\,[\ln(\frac{r_e}{r_w}) - \frac{1}{2}]}{(P_i - P_{wf})} \quad (5\text{-}31)$$

Convert t to days. Then ϕc is calculated from t and t_{Dd} at the matchpoint using the equation

$$\phi c = \frac{t}{t_{Dd}} \frac{.00634\,k}{\mu r_w^2 \frac{1}{2}[(\frac{r_e}{r_w})^2 - 1][\ln(\frac{r_e}{r_w}) - \frac{1}{2}]} \quad (5\text{-}32)$$

5.3.3.1.2 *Empirical Region — find* q_i*,* D_i*, b*

b is obtained from the curve which was matched. q_i is calculated from

$$q_i = \frac{q}{q_{Dd}} \quad (5\text{-}33)$$

D_i is calculated from

$$D = \frac{t_{Dd}}{t} \quad (5\text{-}34)$$

q_i will have the same units as those used to plot q. D_i will have the same units as those used to plot t.

5.3.3.1.3 *Moving* q_i *and* D_i

When the curve has been matched in the empirical region (pseudo-steady state)—which will occur with most decline curves—the values of q_i and D_i will be known as of the first day of production that is plotted on the log-log paper and used in the type curve match. In order to project future production we must know q_i and D_i on the effective date of the study. Finding the new q_i is no problem. Simply use the familiar decline equation.

$$q_{i\,new} = q_{i\,old}\,(1 + bD_i t)^{-\frac{1}{b}} \quad (5\text{-}35)$$

Since D changes with time in hyperbolic decline, it is necessary to calculate a new D_i as of the effective date. This is done using the following equation:

$$D_{i\,new} = \frac{D_{i\,old}}{(1 + bD_{i\,old}\Delta t)} \quad (5\text{-}36)$$

where

Δt = the time from the first day of production to the effective date.

Remember to watch the units on D_i and Δt. Since we are dealing with nominal decline, it is necessary only to multiply by the correct time conversion factor. (Have we convinced you to use nominal decline rates yet?)

Example 5-9
Hyperbolic Decline — Moving q_i and D_i

The q_i and D_i for the beginning of month one, year one, found from the type curve match in Example 5-8 are 3030 BOPM and 0.042/month respectively. If the reserves are to be forecast effective the first month of the fifth year for this well, determine q_i and D_i for this effective date. Remember, b remains constant and has a value of 0.6 for this well.

$$D_i = 0.042/\text{month}$$

$$Q_{i_{1/1}} = 3030 \text{ BOPM}$$

Determine new Q_i:

$$t = 48 \text{ months}$$

$$Q_{i_{1/5}} = Q_{i_{1/1}} (1 + b\, D_i\, t)^{-\frac{1}{b}} \quad (5\text{-}35)$$

$$Q_{i_{1/5}} = 3030(1 + 0.6 * 0.042 * 48)^{-\frac{1}{0.6}}$$

$$= 808.3 \text{ BOPM}$$

Determine new D_i:

$$D_{i_{1/5}} = \frac{D_{i_{1/1}}}{1 + b\, D_{i_{1/1}}\, t} \quad (5\text{-}36)$$

$$= \frac{0.042}{1 + 0.6 * 0.042 * 48}$$

$$= .019/\text{mo.}$$

5.3.3.1.4 *Shifting the Curve*

Occasionally we will have a well with a long production history. Because of the nature of logrithmic curves, the rate-time data will be compressed, and it will appear that the well is declining exponentially. When this occurs, it is helpful to "shift" the curve by redefining time 0. Plot only the last three or four years of production history with the new "first" month's data plotted at .5 months. This will stretch out the curve and enable a more meaningful type curve match. When this is done, remember that q_i and D_i are defined as of the new time 0, and are not for the actual first day of production which may have occurred 20 years before the new time 0.

It is good idea to try to type curve match all decline curves. Type curve matching allows one to make sure that pseudo steady-state has been reached and also gives an idea of how many different curves might fit the same data. Once the data has been satisfactorily matched there is one more step. Always plot the data on a standard semi-log plot along with the "best fit" line. This allows a final check to make sure there are no stupid errors and also allows you to visually check the fit. This step is very important. If you don't like the look of the fit, then you should re-match the data.

5.3.3.2 *Trial and Error*

One of the methods of fitting hyperbolically declining data is trial and error. It is much faster if a calculator has been programmed to calculate q given all of the other parameters.

The steps to follow are:

1. Plot the data on a piece of semi-log paper and draw a smooth line through it.
2. Pick two points on the smooth curve near the extremities of the smooth curve. (t_1, q_1 and t_2, q_2)
3. Guess a value of b.
(A way to get a good starting value is discussed later)

4. Calculate D_i

$$D_i = \frac{(\frac{q_2}{q_1})^{-b} - 1}{b(t_2 - t_1)} \quad (5\text{-}37)$$

This equation can also be programmed into the calculator.

5. Knowing q_i (q_1), D_i, and b, calculate q for a time near the middle of the smooth curve.

6. If the calculated value of q from step 5 falls on the smooth line, calculate more values of q and make sure the smooth line is a hyperbolic curve.

   Otherwise:

   If q falls below the smooth line, you guessed b too large.

   If q falls above the smooth line, you guessed b too small.

   Correct b and go to step 4.

This method sounds tedious, but if a programmable calculator is available, it converges rapidly.

Just as with type curve matching, plot the calculated curve through the data and look at it. Also remember that q_i and D_i are defined at point 1 (t_1). You must "move" them to the effective date using equations 5-35 and 5-36.

5.3.3.2.1 *Guessing 'b'*

A reasonably good estimate of b can be obtained from curves derived by Gentry (1972) (Figure 5-6). Two points are picked from a smooth curve drawn through the data (t_1, q_1) and (t_2, q_2). The cumulative production between t_1 and t_2 is divided by the product of t_2—t_1 and q_1. This result is used to enter the graph on the X-axis. The graph is entered on the Y-axis with q_1 divided by q_2. The value of "b" is read from the graph. In equation form

$$X = \frac{\Delta N_p}{q_i (t_2 - t_1)} \quad (5\text{-}38)$$

$$Y = \frac{q_1}{q_2} \quad (5\text{-}39)$$

It is recommended that this value of b be used in a trial and error calculation as discussed previously.

Example 5-10
Hyperbolic Decline — Trial and Error

For this problem, use the production data for Hyperbolic Well3 from Example 5-8. In the text, follow the algorithm shown in the trial and error section.

1. Plot the data on semilog paper. Draw a smooth line through it.

2. Pick two points *on* the smooth curve and near the extremities of the smooth curve.

   t_1 = 0.5 months (mid point value for first month)
   q_1 = 2978 BOPM
   t_2 = 47.5 months
   q_2 = 825 BOPM

3. Assume a value of b. Let's use the curves derived by Gentry (Figure 5-6) to determine a starting point for b.

   From Figure 5-6

$$X = \frac{\Delta Np}{q_1 (t_2 - t_1)}$$

$$= \frac{74567}{2978 (47.5 - 0.5)}$$

$$= 0.53$$

$$Y = \frac{q_1}{q_2}$$

$$= \frac{2978}{825}$$

$$= 3.61$$

$$b \sim 0.5$$

The assumption of $\Delta Np = 74567$ is high since it is for a full 48 months.

4. Calculate D_i

$$D_i = \frac{(\frac{q_2}{q_1})^{-b} - 1}{b(t_2 - t_1)} \qquad (5\text{-}37)$$

$$D_i = \frac{(\frac{825}{2978})^{-0.5} - 1}{0.5(47.5 - 0.5)}$$

$$= 0.038/\text{mo.}$$

5. Calculate q for time 23.5 months. Refer to Decline Curve Equations at the end of the chapter.

$$q_{23.5\ \text{months}} = q_{0.5\ \text{months}}(1 + bD_i t)^{-\frac{1}{b}}$$

$$= 2978(1 + 0.5 * 0.038 * 23.0)^{-\frac{1}{0.5}}$$

$$= 1442\ \text{BOPM}$$

Since all our production falls on the smooth curve, we can check our calculated value of $q_{23.5\ \text{months}}$ with the tabled production history (see Example 5-8). For month 24 the average production is 1407 BOPM (remember, we assume that the average period rate plotted at the midpoint of the period is a close approximation of the instantaneous rate at that point. That is why we assumed t = 23 months). $\Delta t = 23.5 - 0.5$

6. Although 1407 BOPM from the smooth curve is close to the calculated 1442 BOPM, let's try another value of b. Since the calculated 1442 BOPM is greater than 1407 BOPM on the smooth curve, our next guess for b should be larger. Let's try b = 0.6.

$$D_i = \frac{(\frac{825}{2978})^{-0.6} - 1}{0.5(47.5 - 0.5)}$$

$= 0.041/\text{mo.}$ (Compare to type curve results Example 5-8).

$$q_{23.5} = 2978(1 + 0.6 * 0.041 * 23)^{-\frac{1}{.6}}$$

$$= 1410\ \text{BOPM}$$

The calculated q of 1410 BOPM compares to the value of 1407 BOPM on the smooth curve. It appears then, that a b value of 0.6 is a good assumption. However, other values of q should now be checked to insure that our assumption of $D_i = 0.041/\text{mo.}$ and $b_i = 0.6$ will fit more than just three points on the smooth curve.

For t = 12.5 months (Jan. 19x2)

$$q_{12.5} = 2978(1 + 0.6 * 0.041 * 12)^{-\frac{1}{.6}}$$

$$= 1935.1\ \text{BOPM}$$

This compares to 1932 BOPM.

For t = 30.5 months (July, 19x3)

$$q_{30.5} = 2978(1 + 0.6 * 0.041 * 30)^{-\frac{1}{.6}}$$

$$= 1185.\ \text{BOPM}$$

This compares to 1182 BOPM.

FIGURE 5-5

Determining the value of the hyperbolic exponent "b"

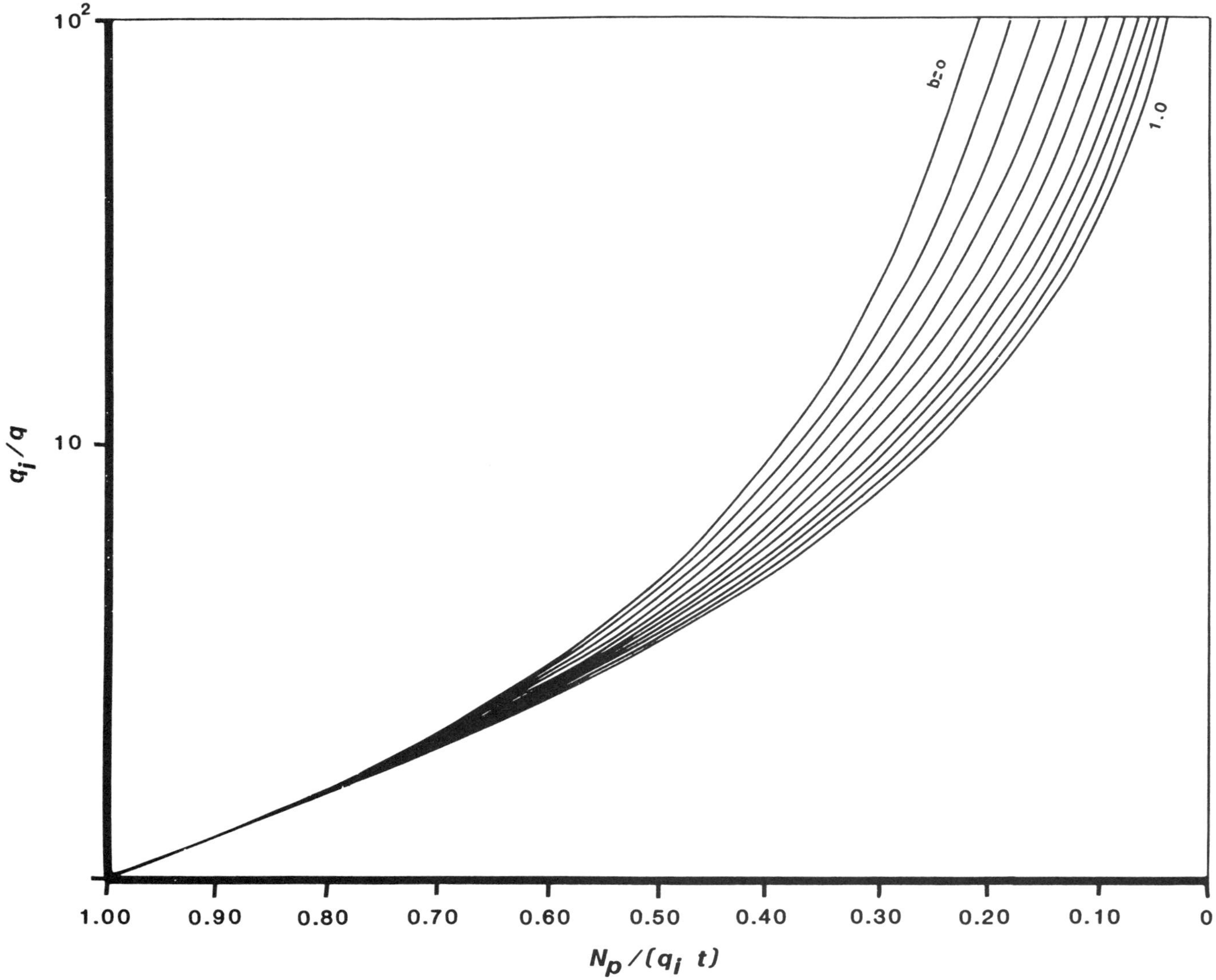

5.3.3.3 *Finding D_i from a Tangent*

Since D_i at any time t is the rate of decline, D_i is the slope of the smooth curve drawn on semilog paper. A tangent to the curve can be drawn by using a mirror to construct a perpendicular to the smooth curve. Then the mirror is used to construct a perpendicular to the perpendicular. This is a line tangent to the smooth curve. Values are read from the tangent and D_i is calculated by

$$D_i = \frac{\ln\left(\frac{q_i}{q_2}\right)}{t} \qquad (5\text{-}40)$$

This is the nominal decline rate at t_1. It is now possible to solve for b knowing t_1, q_1, t_2, q_2 and D_i. The solution is trial and error so it is helpful to have a programmable calculator. An initial guess for the value of b can be made as previously discussed.

5.3.3.4 *Special Graph Paper*

The hyperbolic equation can be expressed as a staight line as follows:

$$q = q_i\,(1 + bD_i\,t)^{-\frac{1}{b}} \qquad (5\text{-}41)$$

$$q^{-b} = q_i^{-b} + q_i^{-b}\,bD_i t \qquad (5\text{-}42)$$

This is of the form

$$Y = b + mx \qquad (5\text{-}43)$$

If q^{-b} is plotted versus time, a straight line should result, *provided* the correct value of b has been chosen. If the assumed b is too small then the line will curve upward. If the assumed b is too large then the line will curve downward. A reasonable starting value for b can be obtained as previously discussed.

It is useful to construct specially scaled graph paper on which to plot the data. The function being plotted is actually q^{-b}. Appendix C illustrates the method of constructing specially scaled paper.

Once a straight line has been obtained, the value of q_i can be read directly. The value of D_i (defined at the same point as q_i) can be calculated from the slope (m) where

$$m = \frac{q_2^{-b} - q_i^{-b}}{t_2 - t_1} \qquad (5\text{-}44)$$

Then

$$D_i = \frac{m}{q_i^{-b}\,b} \qquad (5\text{-}45)$$

The units on D_i will be the same as the units on t. All of the previous comments apply — plot the data on semilog paper and look at it, "move" D_i and q_i to the effective date, and the units on D_i can be changed by multiplying by the appropriate time conversion factor.

5.3.3.5 *Ranges of "b"*

It would be helpful to know what value of "b" to expect under what circumstances. Prior to Fetkovich's paper in 1973, the efforts to determine reasonable values of b primarily consisted of calculating numerous decline curves. In 1944, Arps published an analysis of the data on 149 leases collected by Cutler in 1924. According to Arps' analysis, most of the values of "b" were between 0.0 and 0.4 with none being above 0.7. Ramsay (1968) examined 202 leases and found an approximately uniform distribution of values of b between 0.0 and 1.0. Figures 5-7 and 5-8 contain histograms of Arps' and Ramsay's results.

Additional work was done by Fetkovich (1973b) in an effort to establish values of "b". Fetkovich combined an "Absolute Open Flow" (AOF) equation for oil with two approximations of the material balance equation to create decline curves. The rate equation he used is similar to the familiar "AOF" equation for gas

$$q = J'_{oi}\,\frac{\overline{P}}{P_i}\,(\overline{P}^2 - P_{wf}^2)^n \qquad (5\text{-}46)$$

where

J'_{oi} = the initial productivity index, $STB/D/(psi)^{2n}$

FIGURE 5-6

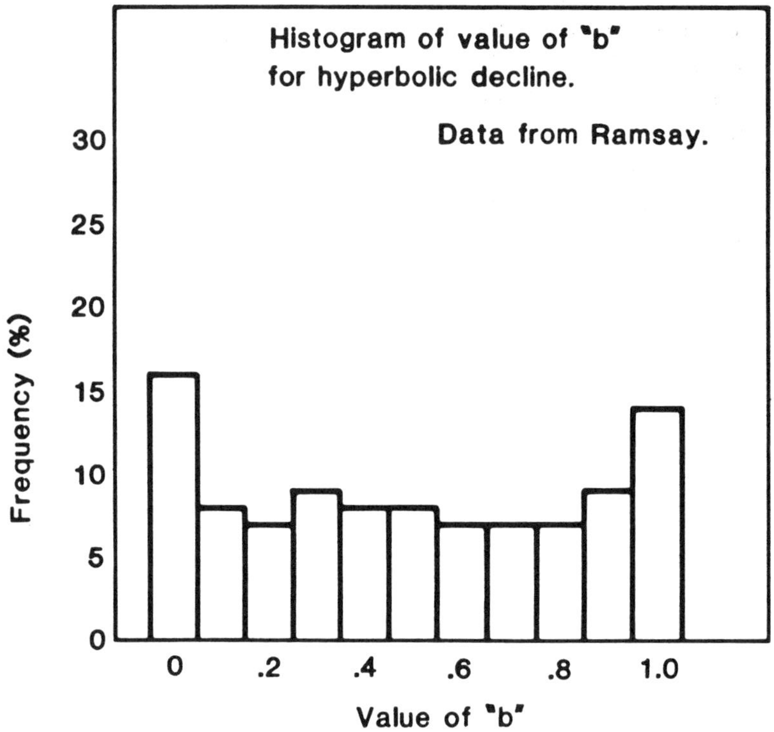

FIGURE 5-7

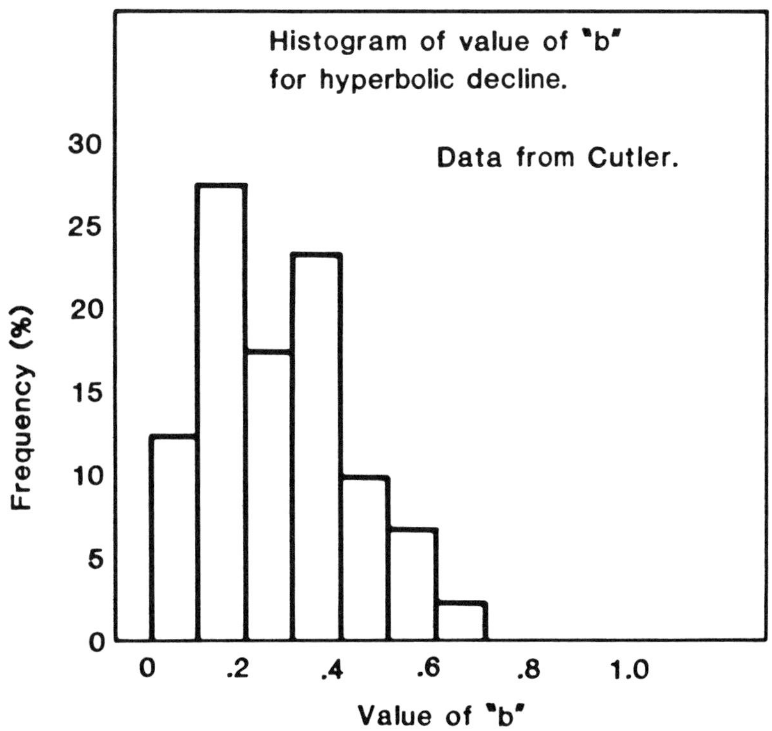

$\overline{P}$ = average reservoir pressure, psi

P_i = initial reservoir pressure, psi

P_{wf} = bottom hole flowing pressure at the well, psi

n = backpressure curve exponent

Fetkovich states that this equation compares favorably with computer results for two phase flow, but it has not been field tested.

In gas well testing, the exponent "n" varies from .5 for fully turbulent flow to 1.0 for laminar flow. This range of n was investigated in equation 5-46.

Two forms of the material balance were investigated.

1) $\overline{P}$ vs N_p is linear.

2) $\overline{P}^2$ vs N_p is linear.

Concept 1 is correct for gas reservoirs if z does not vary with pressure and is a good approximation for oil reservoirs above the bubble point. Concept 2 approximates the shape of a solution-gas-drive reservoir, although concept 1 is a reasonable approximation during much of the decline period.

When the "AOF" equation was combined with the material balance equations, the following values of "b" were obtained

Values of "b"

	n = 0.5	n = 1.0
P vs N_p linear	0.5	0.667
P^2 vs N_p linear	0.5	0.333

Combining the standard gas AOF equation

$$q = C\,(\overline{P}^2 - P_{wf}^2)^n \qquad (5\text{-}47)$$

with the linear P vs G_p form of the material balance resulted in:

Values of "b"

	n = 0.5	n = 1.0
Gas Well	0.0	0.500

Although these values of b are empirical, they do give us an idea of what range of b to expect. The results of the theoretical and empirical work done to establish values of b is summarized below.

b = 0

Oil reservoirs above the bubble point.

Oil reservoirs producing under gravity drainage with no free surface.

Gas reservoirs with constant compressibility (high pressure).

Gas reservoirs in turbulent flow.

b = .333

Oil reservoirs under solution-gas-drive and laminar (?) flow.

b = .5

Oil reservoirs under gravity drainage with a free surface (secondary gas cap).

Oil reservoirs in turbulent (?) flow.

Gas reservoirs in laminar flow.

b = .667

Oil reservoirs where P vs N_p is linear and in laminar (?) flow.

The question marks behind the words laminar and turbulent are to warn the reader that they might not be the right terminology. Fetkovich used $n = .5$ for turbulent flow and $n = 1$ for laminar flow in the "AOF" equation. In gas wells these values of n are associated with turbulent and laminar flow. The extension of these concepts to oil reservoirs may not be correct.

Although the above discussion does not cover all conceivable production schemes, it does shed some light on the expected values of "b". Probably the most important conclusion that can be drawn is that b values above .7 should be viewed with suspicion.

5.3.3.6 *Other Methods*

Other methods exist for calculating decline curves, but they are of little value compared to the ones discussed. Shifting the curves on log-log paper is too tedious and does not lead to reliable answers. Using some sort of least squares regression technique is fraught with danger since the computer usually has no means of rejecting bad data. The extrapolation of decline curves requires good judgment more than it needs calculational speed. Anyone can be trained to do decline curve analysis but the results are not always good.

Another method often mentioned in the literature is the Arps' loss ratio method. As shown in Appendix B, this method has many drawbacks and is definitely *not* recommended.

5.4 SECONDARY STREAMS

The projection of secondary streams is even more difficult than the projection of the primary product. There is no totally satisfactory way of making the projection, but several of the more useful ones will be discussed.

5.4.1 *Projecting Gas Production from an Oil Well*

There are several methods of projecting gas production. Probably the best is using the Tarner material balance technique to predict gas-oil ratio (GOR) versus cumulative oil production. A cumulative oil production versus time curve can be constructed from decline curves and the two curves can be combined to predict GOR versus time. The drawback to this method is that it requires fluid property data and relative permeability data which are often not available.

5.4.1.1 *Constant GOR*

One of the more conservative ways of projecting GOR (early in the life of the well) is to assume it will be constant. Since GOR increases dramatically in a solution gas drive reservoir as the critical gas saturation is reached, this method will predict too low a gas rate. The opposite problem occurs near abandonment. At this time the GOR trends downward so a constant GOR projection may be optimistic.

5.4.1.2 *Log G_p vs Log N_p*

It has been found that the log of cumulative gas production (G_p) versus the log of cumulative oil production (N_p) often plots as a straight line. This information can be used to project gas rate versus time. The equation for a straight line on log-log paper is

$$y = K x^m \tag{5-47}$$

where

m is the slope, inches/inch

K is the value of y at $x = 1$

Once a straight line has been established, measure the slope (rise over run) in inches per inch. Then, using a point on the line, calculate K. You now have the equation.

$$G_p = K(N_p)^m \tag{5-48}$$

In order to calculate yearly gas production, project cumulative oil production as a function of time. Find the cumulative oil production at the end of each year. From equation 5-48 find the cumulative gas production at the end of each year. Subtracting G_p at the beginning of the year from G_p at the end of the year gives ΔG_p or the production for the year.

This method has a couple of interesting features. First, it has been found that about 80% to 85% of the original gas in place will be produced from a solution gas drive reservoir. The original gas in place (G) can be estimated as original solution GOR times original oil-in-place (N). The original oil-in-place is estimated volumetrically. After the straight line has been established, the plot is entered with 80-85% of G to determine an estimate of ultimate oil recovery (EUR). Since the G_p axis runs about one or two log cycles ahead of the N_p axis, a large error in gas in place will result in a small error in oil EUR. This of course assumes that our straight line extrapolation is valid. Caution needs to be used at this point. If a reservoir is above its bubble point pressure, then the producing GOR should be constant. A constant GOR will exhibit a slope of 1 on the log-log plot. Below the bubble point pressure, the GOR will increase and the slope on the log-log plot will also increase. Often the slope will change as the cumulative oil production increases. Thus, although an apparent straight line may be established, the slope may easily change again. Most likely the slope will increase. As the relative permeability ratio of gas to oil (kg/ko) increases, the producing GOR will likely increase. Only near the end of the producing life will the producing GOR decline and thus reflect the decreasing amount of gas in solution. Figure 5-9 demonstrates the problem of establishing a straight line on log-log paper. The first segment has a slope of 1. Likely, we are producing above the bubble point pressure. If the straight line had been extrapolated to determine EUR based on the 80-85% Gp rule, then our EUR would be overstated. In this example there are two apparent changes in the slope.

The second interesting feature of the log-log plot is its ease of use. In order to define the straight line, two points are needed. A convenient set of points are the current cumulatives and the expected ultimate recoveries for oil and gas. This makes it easy to tell a computer program how to schedule production.

No matter what method is used to schedule gas production or GOR, the results should be plotted on the decline curve and visually checked for reasonableness.

5.4.2 *Water-Oil Ratio vs Cumulative Oil.*

In water drive reservoirs or fields under secondary recovery, the amount of water being produced often determines when the well is no longer economic. Oil decline curves often exhibit almost no decline while the water cut in the produced fluid increases. Wells exhibiting this behavior are not amenable to standard decline curve analysis. One empirical method which is sometimes useful is to plot water-oil ratio (WOR) versus N_p. This will sometimes result in a straight line which can be extrapolated to a limiting WOR, usually around 100 to 200 BW per BO. In a situation where the operating costs are sensitive to water production, this method can be used to help project water production and therefore costs. The amount of water produced during a year can be found from the straight line plot and N_p at the beginning and end of the year. The equation of the line is

$$\log \frac{q_w}{q_o} = m\, N_P + b \qquad (5\text{-}49)$$

or

$$\frac{q_w}{q_o} = 10^{(mN_P + b)} \qquad (5\text{-}50)$$

Having found the cumulative oil production (N_p) at the beginning and end of a year (N_{p_1} and N_{p_2}), the amount of water produced during the year (ΔW_p) can be calculated from the average WOR.

$$\text{Average WOR} = 10^{\left(m\left(\frac{N_{P_1} + N_{P_2}}{2}\right) + b\right)} \qquad (5\text{-}51)$$

$$\Delta W_p = \text{Average WOR} * (N_{p_2} - N_{p_1}) \qquad (5\text{-}52)$$

This method can also be used by plotting the log of water cut (WC) versus N_p. Water cut is the ratio of $q_w/(q_o + q_w)$. Since water cut and water-oil ratio are related by:

FIGURE 5-8

Log G_p versus Log N_p

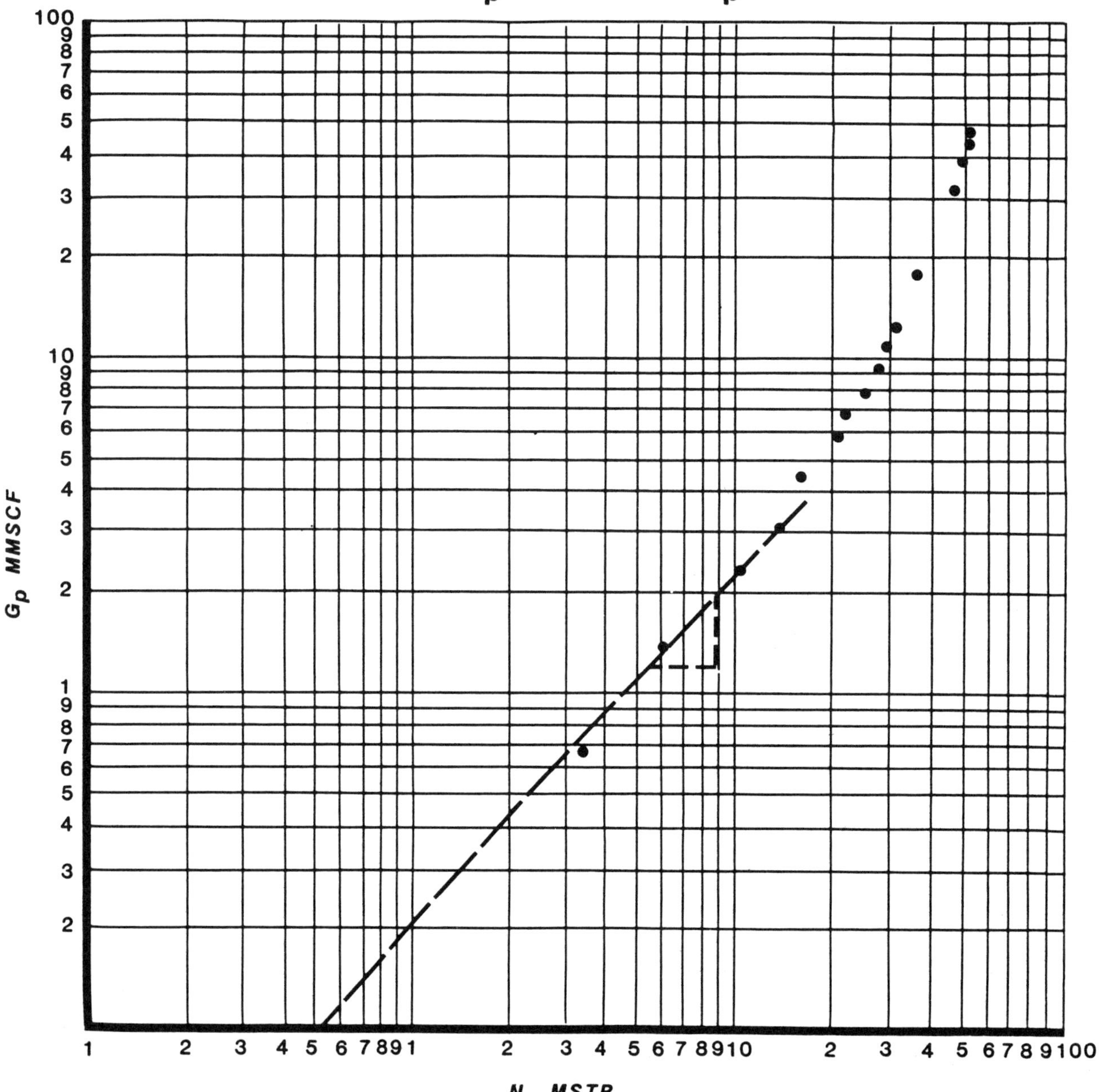

$$WC = \frac{WOR}{1 + WOR} \qquad (5\text{-}53)$$

or

$$WOR = \frac{WC}{1\text{-}WC} \qquad (5\text{-}54)$$

a similar plot results. This plot has the advantage of being bounded since water cut cannot exceed 100%.

5.4.3 *Projecting Condensate*

5.4.3.1 *Constant Yield*

It has been observed that liquid yield (Y, bbls/MMCF) in some gas wells in western Wyoming stays relatively constant or increases slightly with time. In these cases projecting condensate by using a constant liquid yield is reasonable.

5.4.3.2 *Hyperbolically Declining Liquid Yield*

In 1978, Forrest Garb published some empirical equations which showed that liquid yield decreased hyperbolically with reservoir pressure. His equation is

$$Y = Y_i \left[1 + n\, D_i \left(1 - \frac{P}{P_i}\right)\right]^{-1/n} \qquad (5\text{-}55)$$

where

Y_i = initial liquid yield, Bbl/MMCF

Y = liquid yield after some production, Bbl/MMCF

P_i = initial reservoir pressure

P = reservoir pressure after some production

D_i = initial decline rate of liquid yield

n = hyperbolic exponent (b is used in our text)

The values of D_i and n are dependent upon the initial liquid yield.

To calculate "n"

For $Y_i < 15$ Bbl/MMCF

$$n = 24.5 - 1.37\, Y_i \qquad (5\text{-}56)$$

For $Y_i > 15$

$$n = 3.91\, [1 + 2.71\, (Y_i - 15)]^{-0.129} \qquad (5\text{-}57)$$

To calculate "D_i"

For $Y_i < 150$ Bbl/MMCF

$$D_i = 4.05 + .0099\, Y_i^{1.48} \qquad (5\text{-}58)$$

For $Y_i > 150$

$$D_i = 24.3 \ln Y_i - 101 \qquad (5\text{-}59)$$

It can be seen that equation 5-55 is in the same form as the hyperbolically declining rate-time equation.

If the Z-factor is assumed constant, Gp/G can be substituted for the term $(1 - \frac{P}{P_i})$. This results in a more useful equation

$$Y = Y_i \left(1 + n\, D_i \frac{G_P}{G}\right)^{-1/n} \qquad (5\text{-}60)$$

This equation can be integrated to find the cumulative condensate production (N_p) at any cumulative gas production (Gp).

$$Np = \frac{Y_i^{\,n}\, G}{(1 - n) D_i} \left(Y_i^{\,1-n} - Y^{1-n}\right) \qquad (5\text{-}61)$$

where

Np = cumulative condensate produced, Bbls

G = original gas in place, MMCF.

The use of equation 5-61 to find yearly condensate production is very similar to solving the hyperbolic decline equation as shown in Example 5-11.

Example 5-11
Hyperbolically declining liquid yield

Given: $G = 1{,}000$ MMCF

$Y_i = 60$ Bbl/MMCF

Gas Production as shown in following table.

Find yearly condensate production.

1. Find "n"

Since Y_i is greater than 15

$$n = 3.91\ [1 + 2.71\ (Y_i - 15)]^{-0.129} \quad (5\text{-}57)$$

$$n = 3.91\ [1 + 2.71\ (60 - 15)]^{-0.129}$$

$$n = 2.102$$

2. Find "D_i"

Since Y_i is less than 150

$$D_i = 4.05 + .0099\ Y_i^{1.48} \quad (5\text{-}58)$$

$$D_i = 4.05 + .0099\ (60^{1.48})$$

$$D_i = 8.289$$

3. We will find Y at the end of each production increment with equation 5-60 and then calculate cumulative liquid production from equation 5-61.

Year	Annual Gas Volume MMCF	Liquid Yield at End of Year Bbl/MMCF	Cumulative Liquid Production Bbl	Yearly Liquid Production Bbl
0	—	60	—	—
1	230.980	27.84	8,743	8,743
2	162.964	22.49	12,796	4,053
3	114.977	20.19	15,243	2,447
4	81.120	18.95	16,829	1,586
5	57.233	18.20	17,891	1,062
6	40.380	17.72	18,616	725
7	28.489	17.41	19,116	500
8	20.100	17.20	19,464	348
9	13.757	17.06	19,700	236

These equations are probably most useful for Gulf Coast and Mid-Continent production.

The concept behind these equations leads to some interesting possibilities. If Y is plotted vs Gp for a condensate well, the resulting curve should be hyperbolic. If the curve can be fit using any of the techniques shown earlier in this chapter, then one should be able to extrapolate the curve to forecast future production.

5.4.3.3 *Log N_p vs Log G_p Curves*

Just as log Gp vs log Np curves are helpful in predicting gas production for an oil well, log Np vs log Gp curves are useful for predicting condensate from a gas well. The method calculating yearly condensate production from a gas well is the same as that shown for the log Gp vs log Np plot for an oil well.

5.5 SUMMARY

The forecast of future production is the most important single item in an economic evaluation. This chapter has discussed decline curves and other empirical methods of projecting primary and secondary streams. Unfortunately, decline curve analysis is not nearly as easy as it first appears. There are, however, theoretical foundations for its use; hopefully, as the theoretical foundations are developed in future years, decline curve analysis will become more of a science and less of an art.

The following page contains a summary of decline curve equations for your convenience.

DECLINE CURVE EQUATIONS

	Exponential Decline $b = 0$	Hyperbolic Decline $b > 0,\ b \neq 1$	Harmonic Decline $b = 1$
Rate	$q = q_i\, e^{-Dt}$	$q = q_i(1 + bD_i\, t)^{-1/b}$	$q = \dfrac{q_i}{1 + bD_i\, t}$
Cumulative $N_P = f(q_i,\, q)$	$N_P = \dfrac{q_i - q}{D}\, f^*$	$N_P = \dfrac{q_i^{\,b}}{D_i\,(1 - b)}\, [q_i^{\,1-b} - q^{1-b}]\, f^*$	$N_P = \dfrac{q_i}{D_i}\, \ln \dfrac{q_i}{q}\, f^*$
Nominal Decline Factor $D = f\,(D_i,\, t)$	$D = D_i$	$D_i = \dfrac{D_i}{1 + bD_i\, t}$	$D_i = \dfrac{D_i}{1 + D_i\, t}$
Nominal Decline Factor $D = f\,(D_e)$	$D = -\ln\,(1 - D_e)$	$D_i = \dfrac{1}{b}\, [(1 - D_{ei})^{-b} - 1]$	$D_i = \dfrac{D_{ei}}{1 - D_{ei}}$
Life t	$t = \dfrac{\ln\,(q_i / q)}{D}$	$t = \dfrac{(\frac{q_i}{q})^b - 1}{bD_i}$	$t = \dfrac{\frac{q_i}{q} - 1}{D_i}$

$^*f$ = factor to cancel time units (if q is BOPD & D is 1/yr., f = 365)

References

Arps, J.J., 1944, Analysis of Decline Curves: AIME Transactions v. 160, p. 228-247.

Cox, Dave O., 1978, Reservoir Limit Testing Using Production Data: Log Analyst (March-April), p. 13-17.

Cutler, W.W., Jr., 1924, Estimation of Underground Oil Reserves by Oil-Well Production Curves: Department of the Interior, Bull. 228, 114 p.

Earlougher, R.C., Jr., 1977, Advances in Well Test Analy sis: New York, Society of Petroleum Engineers of AIME.

Ehlig-Economides, C. A., and Ramey, H.J., 1979, Transient Rate Decline Analysis for Wells Produced at Constant Pressure: AIME-SPE paper 8387, 10 p.

Fetkovich, M.J., 1973, Decline Curve Analysis Using Type Curves: AIME-SPE paper 4629, 28 p.

Fetkovich, M.J., 1973b, "The Isochronal Testing of Oil Wells": Paper SPE 4529 presented at the 48th Annual Fall Meeting, Las Vegas Nevada.

Garb, F.A., 1978, Liquid Yield for Depletion-Drive Reservoirs and Gas Recovery Factors for Depletion-Drive and Water-Displacement Reservoirs, Petroleum Engineer International, October, 1978, p 72-80.

Gentry, R.W., 1972, Decline Curve Analysis: Journal of Petroleum Technology, January, 1972, p 38-41.

Lewis, J.O., and Beal, C.H., 1918, Some New Methods for Estimating the Future Production of Oil Wells: Trans. AIME, p. 492-525.

Matthews, C.S. and Lefkovits, H.C., 1956, Gravity-Drainage Performance of Depletion-Type Reservoirs: Trans. AIME 207, p. 265.

Ramsay, H.J., 1968, The Ability of Rate-Time Decline Curves to Predict Future Production Rates: Master of Science thesis, The University of Tulsa, 96 p.

Problems

5-1 Calculate the oil in place for a Kansas oil well using the following assumptions.

Drainage area = 40 acres
Formation thickness = 5 feet
ϕ = 0.15
Sw = 0.25
Bo = 1.2

5-2 An oil well is currently producing 100 BOPD. It has a nominal exponential decline rate of 35% per year. At what rate will it be producing two years from now?

5-3 Production data for a well has been plotted on semi-log paper. Careful analysis indicates that the well is declining hyperbolically. Your best fit has a "b" of 0.5, qi of 80 BOPD and a Di of 0.65/year on August 1, 1982. Actual production between July 1, 1982 and 12-31-82 is 12,000 Bbl.

a) What are the flow rate (BOPD), decline rate (1/year), and "b" on 1-1-83?

b) Effective 1-1-83, how many barrels are remaining if the economic limit is 3 BOPD?

c) What are Qi and Di on 7-1-82?

d) Using the hyperbolic fit of the data, what would be the production between 7-1-82 and 12-31-82?

e) What is the EUR for this well (Qel = 3 BOPD)?

5-4 The fit of the production data for a well has the following parameters effective 1-1-83:

Qi = 18250 Bbl/year
Di = .00246/day
b = 0
The economic limit is 5 BOPD

a) How much do you estimate will be produced during 1983?

b) What is the estimated production for 1985

5-5 The rate is 80 BOPD at the beginning of the year and 35 BOPD at the end of the year. The economic limit is 4 BOPD.

a) What is the effective decline rate (1/year) if the production declines exponentially?

b) What is the nominal decline rate (1/year) if the production declines exponentially?

c) What is the EUR for this well (initial rate is 80 BOPD)?

5-6 The following decline parameters are known:

Qi = 400 BOPD
D = 0.5/year
Np = 284,700 Bbl
b = 0

a) What is the economic limit in BOPD?

b) How many years will it take to reach the economic limit?

5-7 The cumulative production to date has been 23,000 STB and 18,400 MCF. The estimated ultimate recovery for oil is 92,000 STB. The EUR for gas is 260,000 MCF. The estimated oil production for next year is 10,000 STB. What is the estimated gas production for the next year? Log Gp vs log Np is a straight line.

5-8 The following rate versus cumulative porduction data is known:

Rate BOPD	Np (Bbl)
50	0
47	2000
39	8000

What is the estimated ultimate recovery to an economic limit of 3 BOPD?

5-9 Log analysis for a Red River C completion yields the following:

$\overline{S_w}$ = 0.40
ϕ = 0.15
h = 5 feet

PVT data:

Pressure	Bo(RB/STB)	Rs(SCF/STB	Relative Volume (V/Vsat)
3300			0.9886
3200			0.9913
3100			0.9942
3000			0.9973
2923	2.083	1635	1.000
2700	1.988	1470	1.0289
2400	1.884	1285	1.0943

Initial reservoir pressure is 3200

Production History:

		Ave. BOPD	Days Prod	BOPM
April	1982	120	30	3600
May	1982	111	31	3441
June	1982	103	30	3090
July	1982	96	31	2976
August	1982	89	31	2759
September	1982	83	30	2490
October	1982	76	31	2356

Economic limit is 3 BOPD

a) If the drainage area is 160 acres, what is the volumetric original oil in place? (Reservoirs of this type usually have a recovery factor of about 20%)

b) Calculate D_i and b as of April 1, 1982

c) Calculate D_i and b as of November 1, 1982

d) Calculate Q for November 1, 1982.

e) How many barrels do you estimate will be produced in November and December of 1982 (total for two months)?

f) Calculate Q for January 1, 1983.

g) What is the estimated ultimate recovery for this well based on decline curves? (3 BOPD economic limit)

h) How many barrels are remaining effective 1-1-83?

i) What is the recovery factor for this well?

j) What is the original gas in place?

5-10 You own a 0.80 working interest and a 0.64 net revenue interest in Lease "A". The current oil price is $30/bbl. If the severance and ad-valorem tax rate is 10%, and the gross monthly operating costs are $2,000:

a) What is the rate at the economic limit in BOPD? (See chapter 8 for economic limit calculation)

b) What must the price of oil be if the economic limit is 2 BOPD

c) If the current rate is 80 BOPD and the decline rate is .5/year (exponential), what % increase in remaining reserves is attributable to this change in economic limit?

5-11 Reporting Year: August 1 to July 31

Fiscal Year BOPD	Average Daily Producton
8/01/79 to 7/31/80	130
8/01/80 to 7/31/81	35
8/01/81 to 7/31/82	30

The best fit of the production is a hyperbolic curve having a Q_i of 200 BOPD, b = 0.5, and Di = 1.50/year on August 1, 1979.

a) If the economic limit is 5 BOPD, what is the estimated month and year of abandonment?

b) How much do you estimate will be produced from August 1, 1982 to abandonment?

c) Oil in place is estimated to be approximately 856,960 barrels of oil. What is the recovery factor for this well?

5-12 Forecast production for each of the calendar years 1983, 1984, 1985 from the following decline curve parameters.

Well put on production 8/1/83
q, D_i, b as of 1/1/84 are 2000 BOPD, 200%/year, .9 respectively
q_{el} = 25 BOPD
1 month = 1/12 year = 30.4 days

5-13 Given:
Di = .47/year
b = 0
N_p = 50,000 bbl
Qel = 2.0 BOPD

a) What is initial rate?

b) What is the life in years?

5-14 Plot the following data:

a)

Date	BOPM
7/83	90
8/83	84
9/83	79
10/83	75
11/83	70
12/83	67

b) Calculate b

c) Calculate D

d) Remaining reserves effective 1/1/84 (Qel = 1 BOPD)

e) Calculate remaining life (from 1/1/84) if Qel = 1 BOPD

5-15

Month	Production	Rate (BOPD)	Cum (Bbl)
1/83	1960	63	1960
2/83	1250	45	3210
3/83	1450	47	4660
4/83	1240	41	5900
5/83	1209	40	7109
6/83	1170	39	8279

a) Plot data and analyze decline curve.

b) What is the nominal decline rate (1/year)

c) If Qel = 3 BOPD, what are the remaining reserves effective 7/1/83?

d) If Qel = 3 BOPD, what is the EUR for this well?

5-16

Pressure (PSIG)	Relative Volume (V/VSAT)
5000	0.9577
4000	0.9721
3000	0.9898
2600	0.9983
2525	1.0000
2513	1.0019
2500	1.0040
2000	1.1163

Relative Volume, V/VSAT, is barrels at indicated pressure per barrel at saturation pressure. B_o at the bubble point pressure is 1.6 RB/STB

a) What is the bubble point pressure?

b) Original reservoir pressure is 4000 psig.
ϕ = .15
h = 20 FT
Sw = .35
A = 80 Acres
How much oil is in place?

c) Is the reservoir saturated or undersaturated?

d) The reservoir pressure drops to 2600 psig. How much oil is produced? Assume a constant volume reservoir. Ignore pore and water compressibility.

e) Would you expect the decline exponent "b" to most likely have a value of (while above the bubble point):
1) b = 0
2) b = .333
3) b = .5
4) b = .667
5) b = 1.0

f) If the production declines exponentially, what is the production rate when the pressure drops to 2600 psig? It is expected that the nominal decline rate will be 0.80/year. Qi = 100 BOPD

5-17 Calculate the apparent drainage area in acres for the following case:

Q_i = 300 BOPD
Qel = 2 BOPD
b = 0.4
Di = 3.0/year
ϕ = 0.13
h = 16 feet
Sw = 0.45
RF = 15%
B_o = 1.8

5-18 Given the following:

	Pressure (Psia)	Bo	Cum Production Bbl
	5000	1.7239	0
	4500	1.7363	2143
	4000	1.7498	5181
	3500	1.7647	9248
	3000	1.7816	14576

Initial Pressure = 5000 psia.
Bubble point pressure = 2700 psia
Assume: Oil compressibility is the only energy mechanism. Ignore pore and water compressibility. The small amount of water production can be ignored.

a. What is the best estimate of oil-in-place using material balance?

b. Calculate oil-in-place using rate-versus-time plot and the following production history.

Q_{el} = 2 BOPD.
Typical recovery factor = 25%.

Date	Q (BOPD)
1/1/83	100
1/1/84	90
1/1/86	70

c. Which answer is probably the best? Defend.

Table of Contents
Chapter 6
Oil and Gas Prices

6.1 Nonregulatory Aspects of Oil Prices 6-1
6.2 Nonregulatory Aspects of Gas Prices 6-2
6.3 Gas-Oil Equivalence 6-2
6.4 Oil Price Regulation 6-3
6.5 Gas Price Regulation 6-4
6.5.1 Gas Prices—NGPA 6-5
6.5.1.1 New Gas Supplies 6-5
6.5.1.2 Existing *Interstate* Contracts 6-6
6.5.1.3 Existing *Intrastate* Contracts 6-6
6.5.1.4 Other Categories 6-7

Chapter 6

Oil and Gas Prices

The single most important variable in oil property evaluation is the determination of reserves and the forecasting of production. Crude oil and gas pricing, at least recently, has also become very important. The engineer must not only solve the difficult task of production forecasting, but he must now devote considerable time to deciphering the myriad of government price regulations. At one time oil pricing was relatively easy, oil was simply called oil and except for differences in quality and transportation we only had one price. However, government intervention in the form of price controls changed this scenario, especially in the 1970's. As a result of the Economic Stablization Act of 1970 oil became old oil, new oil, stripper oil, and released oil. Later, new oil and stripper oil became upper tier oil; and old oil became lower tier oil. And again later, upper tier oil became stripper oil and upper tier oil; while lower tier oil remained lower tier oil. Finally, in June 1979, lower tier oil gradually was reclassified as upper tier oil; upper tier oil was released to market price; stripper oil remained stripper but was permitted market price; and some oil became ''newly discovered''. On January 28, 1981, oil again became oil, with the price determined through market forces. Gas pricing, unfortunately, is more complicated; having at least 20 price categories.

This chapter will present the nonregulatory factors of pricing such as BTU content and API gravity, a brief history of oil and gas pricing, and the current pricing policy.

The summary of federal price regulations is not intended to be the basis for legal advice. Rather, the objective is to give the reader a fundamental basis for understanding the current price structure. Federal regulations and opinions are issued daily. In order to form an opinion on any specific problem, careful analysis of existing opinions, regulations, decisions, and statutes is necessary.

6.1 NONREGULATORY ASPECTS OF OIL PRICES

Actually, this aspect of pricing is relatively simple. Oil prices are influenced by the API gravity and sulfur content of the oil. The API gravity is usually determined by the use of a hydrometer. The scale on the hydrometer may read either specific gravity or °API. First, let's define specific gravity and then define the relationship between specific gravity and °API.

Specific gravity. The specific gravity of a hydrocarbon mixture is the ratio of the mass of the hydrocarbon mixture to the mass of an equal volume of water at some standard temperature. For API gravity the standard temperature is 60°F.

API Gravity. The API gravity or degrees API (°API) is calculated using the following relationship:

$$°API = \frac{141.5}{SG\ 60/60F} - 131.5$$

where:

SG 60/60F = specific gravity of hydrocarbon mixture when the mixture and the equal volume of water are measured at 60°F.

Most often the quantity of oil is measured at a temperature that is not equal to 60 °F. However, crude oil price bulletins published by purchasers give the price of oil in terms of the oil gravity measured at 60 °F. The gravity of the oil at the measured temperature must be corrected to the standard temperature of 60 °F. This is easily done by using tables published by the American Society for Testing Materials (ASTM) although a specific laboratory test for your crude oil may be warranted. When the observed temperature is less than 60 °F, then API gravity at 60 °F will be higher than the API gravity at the observed temperature. The opposite occurs at temperatures greater than 60 °F. Also, the amount of oil sold must be measured at standard conditions. The actual sales volume is less than the measured volume if the oil is measured at temperatures greater than 60 °F and greater than the measured volume for oil measured at temperatures less than 60 °F. Again, tables have been published by the ASTM that give correction factors to adjust the volume of oil at some measured temperature to the standard temperature of 60 °F. The tables assume API gravity is measured at 60 °F so this adjustment needs to be made first.

Often, basic sediment and water (BS&W) is produced along with the oil. BS&W present in the stock tank and trucked or metered in the lease automatic custody transfer (LACT) unit is measured so that the final oil sales volume can be determined. Thermal shrinkage and BS&W may account for as much as seven percent of the original volume. In such cases, decline curve analysis using unadjusted production figures would overstate the reserves.

A crude oil price bulletin that was published by Amoco, is presented as Table 6-1 and shows the effect of gravity and sulfur on the price. A sample calculation is presented in Example 6-1.

6.2 NONREGULATORY ASPECTS OF GAS PRICES

The price of natural gas ignoring federal regulations is determined by the quality of the gas and the terms for delivery established by the gas contract. The contract will usually specify the minimum acceptable heating value for the gas which is usually measured in terms of British thermal units (BTU's) per cubic foot of gas. The usual minimum value is 1,000 BTU's per cubic foot measured at the temperature and pressure specified in the gas contract. The price of the gas may be adjusted upward or downward to reflect the actual BTU content by multiplying the gas price for "1,000 BTU" gas by the ratio of the actual BTU content to 1,000. The maximum water vapor content of the gas is also specified in terms of pounds of water vapor per million cubic feet of gas. The acceptance of the gas by the purchaser is often subject to stipulations concerning the oxygen, carbon dioxide, hydrogen sulfide, and total sulfur content of the gas.

6.3 GAS OIL-EQUIVALENCE

Many times it is convenient to combine oil quantities and gas quantities. Obviously, since oil is measured in terms of barrels and gas is measured in terms of standard cubic feet, we cannot simply

---

Example 6-1
Determine the Crude Oil Price

For the following problems use the Amoco Crude Oil Price Bulletin 83-2 (Table 6-1).

1. Determine the price per barrel of sweet Oklahoma oil with an API gravity of 38°.

   $30.00/bbl for 40 °F API oil
   Gravity adjustment (Column 3) is $0.04/bbl.
   Price per barrel is $29.96/bbl.

2. Determine the price per barrel for sour Oklahoma oil with an API gravity of 38°.

   $29.00/bbl for 40° API oil
   Gravity adjustment (Column 5) is 0.30. (0.015)(20) = 0.30
   Price per barrel is 28.70

---

add the volume of oil in barrels to the volume of gas in standard cubic feet. We would like to be able to calculate the volume of gas (in MCF) that is equivalent to one barrel. Usually the equivalence is based on a BTU equivalence rather than a price equivalence. One barrel of 40 °API oil is usually assumed to be equivalent to 5.7 MM BTU. If gas is assumed to have one million BTU per thousand standard cubic feet, it takes 5.7 MCF to be equivalent to one barrel of oil. The result is usually reported as barrels of oil equivalent (BOE). The equivalence may also be calculated on the basis of prices. If oil is sold for $30 per barrel and gas for $3.00 per MCF, it takes 10 MCF to equal a barrel.

6.4 OIL PRICE REGULATION

Before August 1971, when President Nixon imposed price controls on all commodities, government involvement in the petroleum industry was relatively insignificant when compared to the voluminous and complex regulations that were to follow. Although in late 1973 most price controls were lifted, the price controls on the petroleum industry were retained.

In October 1973, as a result of the Arab oil embargo and the rapid increase in prices by the Organization of Petroleum Exporting Countries (OPEC), a massive and restrictive set of price controls were established that, as will be shown, changed frequently. The basis for the price controls was the Economic Stablization Act of 1970. As a result of the act, we were faced with classifying oil as old oil, new oil, stripper oil and released oil. Wells producing during 1972 produced old oil except any amounts above a specified base production control level (BPCL). Any amount above the BPCL was new oil. For every barrel produced above the BPCL, a barrel of old oil was released from the old oil classification to new oil. Property where the average well produced less than 10 barrels per day now produced stripper oil. Properties that were brought into production after 1972, produced new oil. Old oil prices were initially limited to the price on May 15, 1973 plus $0.35 per barrel, but was increased by $1.00 by the end of 1973 as a result of the Emergency Petroleum Allocation Act (EPAA). New, stripper, and released oil received market prices.

One result of the price controls was deeper involvement of the government in everyday operations. Refiners with adequate supplies of old oil would be in a better position than refiners that needed to rely on new oil, stripper oil, or imports. To eliminate this problem, an entitlements program was established. Simply, refiners that were rich in old oil supply were forced to buy entitlements (bringing their cost up to some national average). Refiners that were poor in the cheaper oil could sell entitlements.

The EPAA was scheduled to expire in February, 1975 but was extended until a new law was passed. This law, called the Energy Policy and Con-

---

Example 6-2
Calculating Barrels of Oil Equivalence

A certain company reports oil reserves of 500,000 barrels and gas reserves of 1 BCF.

1. What are the reserves in barrels of oil equivalent on a BTU basis?

$$\underset{\text{Oil Reserves Bbl}}{500{,}000} + \underset{\text{Oil Equivalent For Gas Bbl}}{\frac{1\ \text{MMMCF}}{5.7\ \text{MCF/bbl}}} = 675{,}439\ \text{BOE}$$

2. If oil sells for $34.00 per barrel and gas sells for $4.00 per MCF, what are the reserves in barrels of oil equivalent on the basis of price.

$$\frac{\frac{\$34.00}{\text{bbl}}}{\frac{\$4.00}{\text{MCF}}} = 8.50\ \frac{\text{MCF}}{\text{Bbl}}$$

$$500{,}000 + \frac{1{,}000{,}000\ \text{MCF}}{8.5\ \frac{\text{MCF}}{\text{Bbl}}} = 617{,}647\ \text{BOE}$$

---

servation Act of 1975 (EPCA) took effect in February, 1976.

The EPCA extended controls to production that had not previously been controlled. Again, the name we assign to oil changes. Oil is now classified as upper tier oil and lower tier oil. Starting in February 1976, the BPCL is the lesser of the average monthly amount of old oil produced and sold during 1975 or the average monthly amount of all oil produced and sold from the property during 1972. Pre 1976 production below the BPCL is classified as lower tier oil. Amounts above the BPCL will now be called upper tier oil. Properties that began producing after 1975 would also produce upper tier oil. Lower tier could sell for its May 15, 1973 posted price plus $1.35 plus inflation and incentive adjustment factors. Upper tier oil could sell for its September 30, 1975 price less $1.32 plus inflation and incentive adjustment factors. Stripper oil and what was called new oil became classified as upper tier oil. Released oil no longer existed. Shortly after the EPCA was enacted, the Energy Conservation and Production Act was enacted and decontrolled stripper oil. Thus we now had essentially a three tier system of prices; lower tier, upper tier, and market. Also included in the market category are imported oil, incremental tertiary oil starting in 1978, and production from the Naval Petroleum Reserves.

Initially the average price of all categories of domestic oil was limited to $7.66 per barrel plus the incentive and inflation adjustments. The total adjustment, however, was limited to 10%. In September 1976 the average domestic price was permitted to increase by 10% and the adjustment factors were disregarded.

A provision in the EPCA, permitted oil prices to be gradually decontrolled beginning June 1, 1979. Full decontrol was scheduled for October 1, 1981. Production from onshore properties not producing in 1978 and from offshore properties leased after December 1978 were now called newly discovered oil and permitted market prices. Upper tier oil was gradually released to market prices.

Heavy oil (°API of 16 or less) was decontrolled on August 17, 1979. Later the API gravity criterion was changed to 20 °API or less effective on December 21, 1979. To qualify as heavy oil, the oil produced and sold during the last month of production prior to July 1979 must meet the gravity requirements.

Effective June 1, 1979 producers could elect to revise their BPCL to reflect the average daily production for the six month period ending March 31, 1979. If the election was made the revised BPCL was reduced by nine percent as of June 1, 1979. After this, the BPCL would be reduced one and one half percent per month through December 31, 1979. Regardless of whether or not the election was made, the BPCL's would be reduced by three percent per month starting in January 1980. Lower tier oil production in excess of the BPCL was sold at upper tier oil prices. On January 28, 1981 the Reagan Administration decontrolled all oil production and once again oil became oil for pricing purposes.

6.5 GAS PRICE REGULATION

The history of gas pricing will be limited to the Natural Gas Act of 1938 and the current Natural Gas Policy Act (NGPA) of 1978. Because of the numerous details only an overview will be presented. Specific gas pricing questions should be investigated on an individual basis to ensure that the appropriate price regulation is applied. Unfortunately, the numerous price categories of the NGPA has increased the paper work and added confusion to the evaluation process.

Before passage of the NGPA on November 9, 1978, prices of interstate gas were regulated while intrastate gas sales were at market prices. Accordingly, new gas supplies for the interstate market became limited. On the other hand, supplies for intrastate sales became abundant. By 1976, interstate customers were being curtailed. At the same time, intrastate markets had excess deliverability. The interstate-intrastate price dichotomy and the associated supply problem is the primary reason that congress enacted the 1978 NGPA.

The ostensible objective of the NGPA was to:

(1) encourage exploration for new gas supplies, and

(2) deregulate natural gas prices.

The NGPA eliminated the price dichotomy which was one means of encouraging exploration for gas. The abundant intrastate gas could now be sold in the interstate market. However, instead of all gas being deregulated, intrastate gas became regulated. And, unfortunately, one regulated price was not established. Rather, over twenty price categories were created. Some categories will be deregulated and reach market level prices by 1985 and 1987, and some categories will never be deregulated. Furthermore, the act provides that controls can be extended another eighteen months after deregulation.

The following paragraphs will summarize the various price categories established under the 1978 NGPA.

6.5.1 *Gas Prices — NGPA*

Over twenty categories have been established under the NGPA of 1978. Three general price categories are addressed:

(1) new gas supplies.

(2) existing *intrastate* gas contracts.

(3) existing *interstate* gas contracts.

The NGPA establishes ceiling prices for several categories of gas. The ceiling price is based on a specific amount in dollars per MMBTU for the month of April 1977. This base period amount is then escalated to determine the ceiling price in subsequent periods. The escalation may be comprised of two elements.

(1) An annual inflation adjustment factor based on the gross national product (GNP) implicit price deflator plus 0.2 percentage points, and

(2) A real growth factor.

The real growth factor is 3.5% for April 20, 1977 through April 30, 1981 and 4.0% starting May 1, 1981. The established ceiling prices are only maximums and do not dictate that the actual selling price must be that high.

6.5.1.1 *New Gas Supplies*

Three sections of the NGPA addressed the issue of pricing new gas; Sections 102, 103, and 107.

Section 102 regulates prices for new natural gas produced from *new onshore wells, new reservoirs* and *new Outer Continental Shelf (OCS) leases*. The ceiling price for new natural gas started at $1.75/MMBTU on April 20, 1977 and escalated at the rate of inflation (inflation adjustment factor) plus 3.5% (real growth factor) per year until May 1, 1981. After May 1, 1981 until December 31, 1984 the ceiling price escalates at the rate of inflation plus 4% per year. Prices were deregulated on January 1, 1985. The definition of what is ''new'' is discussed in the following paragraphs.

New Onshore Well:

A new onshore well for section 102 gas pricing is:

(1) A well drilled on or after February 19, 1977 and located at least 2.5 miles from the nearest marker well.

(2) A well drilled on or after February, 19, 1977 that is located within the 2.5 mile radius of a marker well *and* is completed at least 1,000 feet deeper than the deepest marker well within a 2.5 mile radius.

A *new* well is any well where the drilling commenced on or after February 19, 1977 or any well deepened after February 19, 1977 by at least 1,000 feet. A *marker* well is defined as any well from which natural gas was produced in commercial quantities any time after January 1, 1970 and before April 20, 1977. The Act provides specific methods for determining the distance between a marker well and a new well.

New Onshore Reservoir:

New natural gas is defined as gas produced from a reservor that did not produce commercial quantities before April 20, 1977. This gas may be either from an old producing well such as behind-the-pipe gas or from a well where suitable delivery facilities were not in existence on April 20, 1977 and gas was not withheld.

New OCS Leases:

A new OCS lease is any lease entered into on or after April 20, 1977.

Section 103 addresses *new* onshore production wells. The ceiling price for Section 103 gas is based on $1.75/MMBTU for April 1977. The ceiling price is to be escalated by the annual inflation adjustment factor. This differs from Section 102 gas in that the ceiling price for section 103 gas is not increased by the real growth factor. Essentially these wells are development wells for reserves already identified. The well must be a new well that was commenced on or after February 19, 1977. The well must not be located on the Outer Continental Shelf, and must satisfy federal and state well spacing requirements. Certain limitations also apply which exclude a new well if it is within a proration unit.

The depth of a well qualifying for new onshore production must also be considered. Prices for new onshore production were deregulated on January 1, 1985 *if* the well is *more* than 5000 feet deep. For new onshore production wells less than 5000 feet deep, a special ceiling price that is midway between the new natural gas ceiling price (section 102) and the new onshore production well ceiling price (section 103) in effect prior to January 1, 1985 will prevail until June 30, 1987.

Section 107, high cost natural gas, is gas:

- produced from geopressured brine;
- produced from a well that is completed below 15,000 feet and where the drilling commenced after February 19, 1977;
- produced from coal seams;
- produced from Devonian shale;
- that FERC ascertains that additional incentives are needed.

The last category is obviously very broad. Included in this category are tight gas formations that have been designated by the FERC. Tight gas formations, if they qualify, have a ceiling price of 200% of the Section 103 price.

The other categories initially had a ceiling price equivalent to new natural gas but have subsequently been deregulated. Deregulation occurred when FERC issued its incremental pricing regulations.

6.5.1.2 *Existing Interstate Contracts*

Sections 104 and 106a cover the various gas vintages that were in existence on April 20, 1977. There are several categories of Section 104 gas which includes gas under contract on April 20, 1977 where the contract is still in effect. The ceiling price varies from $0.39 to $1.63 per MMBTU for December 1978, depending upon the vintage of the production. Section 104 ceiling prices are escalated by the annual inflation adjustment factor. Section 104 gas under the NGPA will *not* be deregulated.

Section 106a deals with the price regulation of interstate rollover contracts. A rollover contract is a contract that existed prior to the enactment date of the NGPA and subsequently expires. For gas committed to the interstate market, the ceiling price is the greater of either $0.54/MMBTU (on April 1977) or any just and reasonable rate established by FERC. Both amounts are adjusted by the annual inflation adjustment factor. Section 106a gas will *not* be deregulated.

6.5.1.3 *Existing Intrastate Contracts*

Sections 105 and 106b specify ceiling prices for gas that was under contract for intrastate sales prior to the enactment of the NGPA. Section 105 is for gas sold under contracts established before enactment of the NGPA and still in effect. Section 106b addresses the issue of gas sold under an intrastate rollover contract.

For section 105 gas two cases must be addressed: (1) on 11/8/78 the contract price exceeds the ceiling price for new natural gas, and (2) on 11/8/78 the ceiling price for new natural gas exceeds the contract price.

1) Contract price greater than new gas ceiling price:

> The ceiling price is the lower of the contract price *or* the contract price on 11/8/78 escalated by the annual inflation adjustment factor until exceeded by the ceiling price for new natural gas (increased by the annual inflation adjustment factor and the real growth factor). At that point the ceiling price becomes the new gas ceiling price.

2) New gas ceiling price greater than contract price:

> At a given point in time, the ceiling price is the lower of the ceiling price for new natural gas or the escalated contract price.

Gas in the above two categories was deregulated on 1/1/85 *if* the contract price on 12/31/84 is $1.00/MMBTU or more and an indefinite price escalation clause does not exist. If the contract price on 12/31/84 is less than $1.00 per MMBTU the gas is not deregulated but is increased by the inflation adjustment factor.

In the case of rollover of intrastate contracts (section 106b) the ceiling price is the greater of: (1) the expired contract price subsequently adjusted by the annual inflation adjustment factor, or (2) $1.00/MMBTU on April 1977 adjusted by the annual inflation adjustment factor.

6.5.1.4 *Other Categories*

In addition to the general categories addressed, there are two sections , 108 and 109 , that cover stripper well natural gas and any gas not covered by any other section of the Act respectively. Stripper well natural gas is nonassociated gas production where the average production for the preceding 90 day period is less than 60 MCF per day. The well must flow at its maximum efficient rate during this period. Production is permitted to increase above the 60 MCF per day limit and still retain stripper well natural gas status if the increase is the result of a "recognized enhanced recovery technique." The ceiling price for stripper well natural gas is $2.09/MMBTU in May 1978 escalated by the same factor used for new natural gas which includes the annual inflation adjustment factor and the real growth factor. Stripper well natural gas will not be deregulated.

Section 109 gas covers all gas not specifically covered by one of the other sections. The Act specifies that the ceiling price for a gas should be the highest ceiling price available if more than one ceiling price could be used.

The ceiling price for section 109 gas is based on $1.45/MMBTU on April 1977 escalated by the annual inflation adjustment factor. Section 109 gas will *not* be deregulated.

A summary of the various pricing categories appears on the following page.

Summary
Natural Gas Policy Act

New Gas Supplies:

Section 102
- New Onshore Wells
- New Reservoirs
- New Outer Continental Shelf Leases

Section 103
- New Onshore Production Wells

Section 107
- High Cost Natural Gas
  1. Designated Tight Gas Formations
  2. Coal Seams
  3. Devonian Shale

Existing Interstate Contracts:

Section 104
- Various gas vintages that were in existence on April 20, 1977

Section 106a
- Interstate Rollover Contracts

Existing Intrastate Contracts:

Section 105
- Gas under Existing Intrastate Contract as of Nov. 8, 1978

Section 106b
- Intrastate Rollover Contracts

Other Categories

Section 108
- Stripper well natural gas

Section 109
- Gas not covered by any other section

AMOCO

Table 6-1

Amoco Production Company
200 East Randolph Drive
P.O. Box 5340A
Chicago, Illinois 60680

February 17, 1983

"Printed with permission of Amoco Production Company."

Crude Oil Price Bulletin 83-2

(Supersedes Crude Oil Price Bulletin 83-1)

This bulletin shows prices and gravity adjustments posted by Amoco Production Company (Amoco) for purchases of crude oil and condensate. Prices are based on the use of 100% tank tables or mutually acceptable automatic measuring equipment with customary adjustment of volume and gravity for temperature and full deduction for basic sediment and water.

Subject to change without notice, Amoco will pay the prices shown for merchantable crude oil and condensate delivered for its account into the facilities of its authorized receiving agency. Merchantable crude and condensate is defined as virgin crude and/or condensate produced from wells which is free of injected or outside foreign contamination, added chemicals containing but not limited to halogenated organic compounds and oxygenated compounds and which is fit for normal refinery processing. Seller warrants that the crude oil and condensate delivered to Amoco shall be of merchantable quality as defined above and fit for normal refinery use. Prices are shown for 40 gravity and above, except as noted.

All prices may be subject to deductions for trucking and other charges where applicable. In the event government regulations require adjustment to Amoco's prices or effective dates, Amoco reserves the right to amend these prices and recover any excess payments by withholding payments from future settlements or by separate invoicing.

Area			Effective February 17, 1983 — Price: $/Barrel	For Gravity Adjustment See Column
Colorado	Western		29.00*	5
	Eastern		30.00*	1
Kansas	Sweet		29.00*	5
Louisiana	Southern Louisiana Light		30.35*	3
	Hackberry (Flat Price)		30.00*	None
	Port Hudson (Flat Price)		30.50*	None
Nebraska	West Panhandle Sweet		30.00*	1
	Sleepy Hollow Area		26.50*	5
North Dakota	Incl. Sheridan, Roosevelt, Richland Counties, Mont.		30.00*	1
	Fryburg-Medora (Flat Price)		28.85*	None
Oklahoma	Sweet		30.00*	3
	Sour		29.00*	5
Texas	East	— Asphaltic (32° Top Price)	26.00*	4
		— East Texas Field (Flat Price)	30.00*	None
		— Quitman Light	28.00*	5
	Gulf Coast	— South Houston (Flat Price)	30.00*	None
		— Hastings (Flat Price)	30.00*	None
		— Willamar (Flat Price)	30.00*	None
		— Fairbanks Area, Gillock, High Island	30.00*	1
	North	—	30.00*	5
	West Central	— Including King & Knox Counties	30.00*	5
	West Texas & New Mexico	— Intermediate	30.00*	1
		— Sour	29.00*	5
Wyoming	Sweet		30.00*	1
	Sour (including Montana fields)		28.00*	2

* indicates change

Table 6-1
(Continued)

Gravity Adjustments to be Deducted from Top Price Posting, $/Barrel

Gravity, °API	1	2	3	4	5
40 and Above	0.00	0.00	0.00	No Deduction to 32.0	0.00
39 to 39.9	0.02	0.02	0.02		Below 40.0 Deduct $0.015 per 0.1 °API
38 to 38.9	0.04	0.04	0.04		
37 to 37.9	0.06	0.06	0.06		
36 to 36.9	0.08	0.08	0.08		
35 to 35.9	0.10	0.12	0.10		
34 to 34.9	0.12	0.16	Below 35.0 Deduct $0.015 per 0.1 °API		
33 to 33.9	0.14	34.0 to 20.0 Deduct $0.03 per 0.1 °API			
32 to 32.9	0.16				
31 to 31.9	0.18			32.0 to 20.0 Deduct $0.03 per 0.1 °API	
30 to 30.9	0.20				
29 to 29.9	0.22				
28 to 28.9	0.24				
27 to 27.9	0.26				
26 to 26.9	0.28				
25 to 25.9	0.30				
24 to 24.9	0.32				
23 to 23.9	0.34				
22 to 22.9	0.36				
21 to 21.9	0.38				
20 to 20.9	0.40				
Below 20	0.42	Below 20.0 Deduct $0.07 per 0.1 °API		Below 20.0 Deduct $0.07 per 0.1 °API	
Examples					
36.4	0.08	0.08	0.08	0.00	0.54
27.6	0.26	2.08	1.21	1.32	1.86
18.3	0.42	5.55	2.605	4.79	3.255

Requests for additional copies of this bulletin, changes in address, or further information regarding any of the prices shown should be addressed to:

Mgr. Crude Oil Supply Negotiations (N.A.)
Amoco Production Company
200 East Randolph Drive M.C. 1307
P.O. Box 5340A
Chicago, Illinois 60680

Field Office Contacts

Area	Address	Phone
North Dakota, Rocky Mountains, D J Basin, Mid-Continent	Amoco Building - Suite 3330 1670 Broadway Denver, Colorado 80202	303-830-4472 303-830-5715
Oklahoma	Lakepoint West Building Suite 180 4045 N.W. 64th Street Oklahoma City, Oklahoma 73116	405-840-2673
Texas (excluding West Texas), Gulf Coast	1919 Smith - Suite 1260 Houston, Texas 77002	713-951-0140
West Texas, New Mexico	P.O. Box 1725 Midland, Texas 79702	915-684-4471
Louisiana, Other Southeastern States	P.O. Box 50879 New Orleans, Louisiana 70160	504-586-6439

Table of Contents
Chapter 7
Oil Field Deals

7.1 **Oil and Gas Rights** 7-1
7.1.1 Mineral Interest 7-1
7.1.2 Leasehold Interest 7-1
7.1.2.1 Primary Term 7-3
7.1.2.2 Lease Bonus 7-3
7.1.2.3 Royalty Interest 7-3
7.1.2.4 Delay Rental 7-3
7.1.2.5 Shut-in Gas Well Clause 7-3
7.1.2.6 Cessation of Production Clause 7-3
7.1.2.7 Pooling/Unitization Clause 7-3
7.1.3 Assigned Rights 7-4
7.1.3.1 Overriding Royalty Interest 7-4
7.1.3.2 Reversionary Interest 7-4
7.1.3.3 Carried Interest 7-4
7.1.3.4 Net Profits Interest 7-5
7.1.3.5 Production Payment 7-5
7.1.4 Operating Interest 7-5
7.1.5 Nonoperating Interest 7-5
7.2 **Operating Agreements** 7-5
7.2.1 Cost and Revenue Sharing 7-6
7.2.2 Operator Responsibilities 7-6
7.2.3 Expenditure Limits 7-6
7.2.4 Non-Consent Clause 7-6
7.2.5 Accounting Procedures 7-6
7.3 **A Technique for Determining WI and NRI** 7-6
7.3.1 The Tree 7-8
7.3.2 The Flowchart 7-9
7.3.3 Table of WI's and NRI's 7-10
7.3.4 Phases of a Deal 7-11
7.3.5 Complex Deal 7-11

Chapter 7

Oil Field Deals

The terms working interest (WI) and net revenue interest (NRI) have been used previously in the discussion on the cash flow diagram for an oil property. This chapter will develop those and other concepts and illustrate how to determine WI and NRI for various deals.

7.1. OIL AND GAS RIGHTS

We normally get the right to explore for and extract oil and gas by leasing that right from the owner of the mineral rights on a piece of property. The mineral interest may or may not be owned by the same person who owns the surface rights. A mineral interest may be bought and sold separately from the surface rights in all states except Louisiana. When this occurs, the minerals are said to be "severed". When the mineral interest is sold it may be sold forever or only for a specific time period. In Louisiana, a mineral servitude may be sold. This is equivalent to a mineral interest except that, by law, it automatically reverts back to the surface owner after 10 years of nonuse.

The term "landowner" is often used to mean the owner of the mineral interest. A landowner having the mineral rights may grant or retain three interests associated with the minerals; a mineral interest, a leasehold interest, and a royalty interest. These rights to oil and gas and various other rights generated in oil and gas deals are discussed in the following paragraphs.

7.1.1. *Mineral Interest.*

The holder of a mineral interest owns the mineral rights (or a portion of them) under a piece of property. As mentioned above, he may or may not own the surface of the ground. Whether the mineral interest holder owns the surface or not, he has the right to reasonable access to develop the minerals. He must however, compensate the surface owner for any damage caused by operations to explore for or produce the minerals.

The mineral interest owner may own all the minerals under a piece of ground or he may only own certain minerals. For instance, one person may own the coal rights, and another may own the oil and gas rights. He may also own an "undivided" partial interest in the minerals. For example, a 50% undivided interest in the minerals under a section means a 1/2 ownership of the minerals under **all** of the section. Contrast this to owning 100% of the minerals under the north half of the section (a divided 1/2 interest).

The owner of the mineral interest may grant a leasehold interest. In that instance the mineral interest owner becomes the lessor, and the owner of the leasehold interest is the lessee. The lessor usually retains a royalty interest as compensation for allowing someone else to extract the minerals in place.

7.1.2. *Leasehold Interest.*

A leasehold interest is conveyed by the mineral interest owner (lessor) to the lessee. The leasehold interest owner, often called the working interest owner, has the right to explore for and produce minerals. A 100% working interest owner would pay all of the costs to develop and produce the mineral and would receive all the revenue less any royalty interest retained by the mineral interest owner in the leasing transaction (See Figure 7-1). Since the leasehold interest is created by a legal

FIGURE 7-1

WI and NRI Terminology Used in Oil Property Evaluation

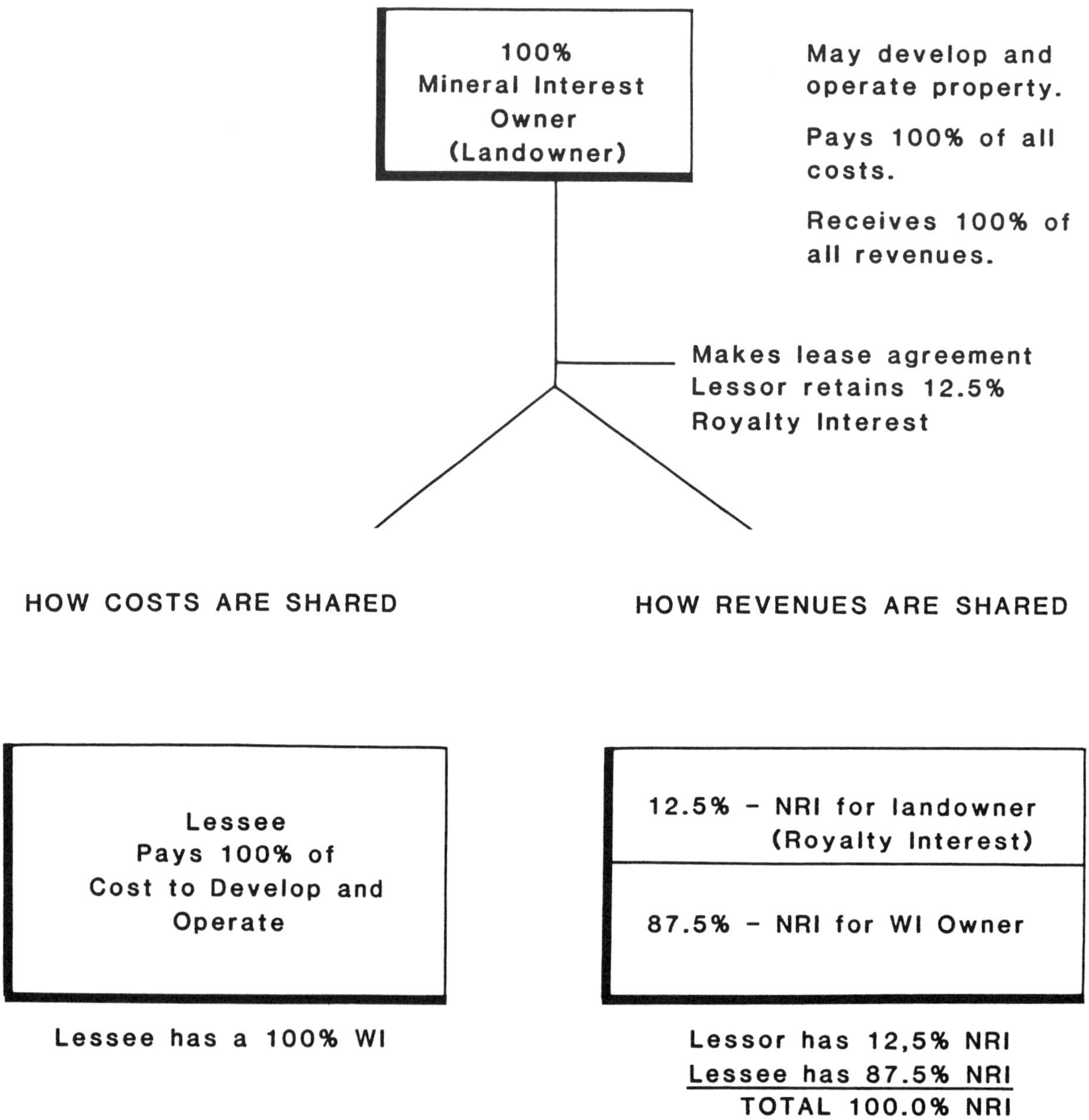

agreement (a lease), it is subject to any provisions contained in the agreement. Common lease provisions are: The primary term, lease bonus, royalty interest, delay rentals, shut-in gas well provisions, pooling and cessation of production clauses.

7.1.2.1. *Primary Term*

Leases are usually in effect for a period of 1 to 10 years (called the primary term) and as long thereafter as hydrocarbons are produced. When negotiating for a lease, the landowner (mineral interest owner) usually tries to get as short a primary term as possible. The lessee wants as long a term as possible since this gives him more time to decide where to drill.

7.1.2.2. *Lease Bonus*

The lessee pays a "sign-up" bonus to the lessor in order to induce him to sign. The amount ranges from $1 per acre for goat pasture to over $3,000 per acre for very good property.

7.1.2.3. *Royalty Interest*

In addition to the bonus, the lessee agrees to pay the lessor a certain percentage of the money received from the sale of hydrocarbons. The lessor does not have to pay any of the costs of drilling and equipping the wells. The lessor also does not pay any of the costs of treating the hydrocarbons except under certain conditions (typically compression, treating of, or transportation of gas). The lessor *does* have to pay his own severance, advalorem, and windfall profit tax.

In the past, the amount of the royalty was almost always 1/8 of the gross proceeds. Now it is not unusual to see royalties of 3/16 or 1/4. In certain areas which are "known" to contain hydrocarbons, the federal government has a sliding scale royalty.

7.1.2.4. *Delay Rental*

Unless operations are commenced on or before one year from the effective date of the lease, the lease terminates (no matter how long the primary term) unless the lessee pays a delay rental. This delay rental allows the lessee to defer the commencement of operations for one year. At the end of that year the lessor can again pay the delay rental and hold the lease. This process continues until the primary term expires or production is established. Typical delay rentals are $1/acre although the lessor can put pressure on the lessee to drill by providing for increasing delay rentals.

7.1.2.5. *Shut-in Gas Well Clause*

Many leases allow the lessee to hold the lease beyond the primary term by paying an amount equal to the annual delay rental if there is a completed gas well on the lease (or on acreage pooled with the lease) which is shut in. This provision is ostensibly to allow the lessee time to develop enough gas reserves to justify a pipeline. In practice it can be used to hold a lease indefinitely at little additional cost if a marginal gas well is drilled. The lessor can thwart this by having large delay rentals or by limiting the amount of time that the shut-in royalty will hold the lease. Two to five years beyond the primary term is reasonable.

7.1.2.6. *Cessation of Production Clause*

Leases normally provide that if a well becomes incapable of production the lessee must commence reworking or drilling operations within 60 days or lose the lease (if the primary term has expired). This can become critical if the only well on a lease goes down for two months. The lease could be lost.

7.1.2.7. *Pooling/Unitization Clause*

Pooling and unitization, strictly defined, have different meanings. Pooling applies to the aggregation of small areas of land necessary to form a drilling unit (acreage needed for one well) as defined by state regulations. Unitization, however, applies to a larger area. Usually the unitized area includes the entire reservoir. Consolidation is necessary for efficient secondary recovery operations such as a waterflood. Without specific pooling or unitization language in the lease agreement, the lessee may not commit his leased area to a unit or pool his leased acreage without the consent of the royalty owner. A pooled or unitized lease will remain in effect as long as the unit remains in effect. Royalty payments are made based on the way production is allocated in the unit.

7.1.3. *Assigned Rights*

The owner of the leasehold interest (or working interest owner) may convey or assign rights provided for by the lease agreement. These rights terminate when the lease expires. Overriding royalty interests, carried interests, reversionary interests, and net profits interests are some examples discussed in the following paragraphs.

7.1.3.1. *Overriding Royalty Interest.*

An overriding royalty or "override" is similar to a landowner's royalty in that the holder of an override receives a portion of the production free and clear of all costs. An override is often said to be carved out of the working interest since it is created from the leasehold's share of the revenue. An override typically arises from the actions of a lease broker who obtains a lease from a landowner and then "turns it" or sells it to an oil company retaining an overriding royalty (typically 1 to 5 percent of 100%). Farmout agreements, probably the most common type of agreement between oil companies, often provide for an override before payout convertible to a working interest after payout. The leasehold interest owner (farmor) of a block of acreage makes a deal (farms out) with another company (the farmee) whereby the farmee will drill a well at its sole risk in return for an interest in a block of acreage. Typically the farmor will retain a 5% override on the initial well until it pays out. At that time the farmor has the option to convert his override to a 25% working interest in the well. Ordinarily, all subsequent wells are paid for on a 75-25 percent basis whether or not the first well has paid out at the time of drilling the subsequent wells.

The amount of the override and back-in will vary, of course. 7-1/2% overrides and 50% back-ins were not unusual during the great turmoil (1977-1981). Sometimes, if the block of acreage is large, there will be a multiple well commitment instead of a single well commitment.

7.1.3.2. *Reversionary Interest*

In the above farmout, the farmor is said to have a reversionary interest. His override *reverts* to a working interest at some specified point (usually payout). A *back-in* is essentially the same as reversionary interest. The farmor might have a 25% backin on the acreage. The best way to describe a reversionary interest or back-in is "a 5% overriding royalty convertible to a 25% working interest at payout, **proportionately reduced**." All deals are assumed to be proportionately reduced unless it is specifically stated otherwise. If a deal is burdened with an overriding royalty of 5% and I take one-forth of the deal, I only pay my proportionate share of the royalty. In this case I pay 25% of 5% or 1.25% of 8/8. The section on determining working intersts and net revenue interest should clarify the back-in or reversion and proportionately reduced concepts.

7.1.3.3. *Carried Interest*

An interest which is not responsible for any costs and does not receive any revenue until a certain condition is met. A farmout could be expressed as a carried interest. Carried interests, however, are more likely to be a carry to the casing point of the first well or through the tanks on the first well (rather than to payout). If a company has a 25% carry to the casing point, then the carrying party pays all of the dry hole costs. In this case dry hole costs are the cost to drill, log, and test the well prior to setting the casing. At the casing point, a carried party having a 25% carried interest has the option of putting up 25% of the completion costs to have a 25% working interest (with its associated net revenue interest) in the well and the other acreage. For a non-producer, dry hole costs would also include the cost to abandon the well.

This is a common way for promoters to get a risk-free look at a play. If the well is a good one they may even be able to borrow the completion money.

A *1/3 for 1/4* is a farm out agreement sometimes referred to as a standard Rocky Mountain deal. The farmee pays 1/3 of the costs to some predetermined point in order to receive a 25% WI. Usually the reversion point is payout. This is the same deal as a 25% back-in after payout. Usually there is an override before payout which converts to a 25% working interest. Notice that if 3/3 of the deal are sold then the farmor pays nothing and gets a 25% WI after payout. This is a pretty standard deal.

7.1.3.4. *Net Profits Interest*

A net profits interest receives a portion of the *net* proceeds from a well after all costs, except windfall profits tax (WPT), have been paid. Usually the costs are cumulative. That is, the net profits interest does not receive any part of the proceeds until payout occurs.

Working interests and net profits interests are treated differently for the windfall profits tax. For tax calculations, net profits interests are "nonoperating" interests and treated the same as royalties (see 7.1.5).

A net profits interest appears quite similar to a carried interest or a back-in. The major difference is that a carried interest or back-in shares in the costs and receipts after specified conditions have been met (casing point, payout, etc.) while a net profits interest usually continues for the duration of production. One party continues to bear costs and the other receives a share of the proceeds after the payment of those costs.

A net profits interest will end up with the same cumulative net cash flow from a successful well as the identical percentage working interest (neglecting WPT). On an unsuccessful well, however, a working interest will lose money while a net profits interest will not share in the losses. You might think of it as a working interest without risk.

7.1.3.5. *Production Payment*

A production payment is a share of the hydrocarbons produced from the lease, free and clear of the costs of production, terminating when a given amount of hydrocarbon has been produced or when a given amount of money from the sale of that hydrocarbon has been realized. It is often used to pay down debt since the amount of money going for debt service is proportional to the producing rate.

7.1.4. *Operating Interest*

A term used by land and legal departments and by the IRS. The operating interest for IRS purposes is any "interest in minerals in place that is burdened with the cost of development and operation of the property" (Burke and Bowhay, 1983). The operator of a well, on the other hand, is the entity which is responsible for the day to day management activities. The operator prepares cost estimates, gets approvals from nonoperators, contracts with the drilling company, supervises the drilling of the well, designs and supervises the completion of the well and does all the accounting. He is reimbursed for his costs by the nonoperators.

7.1.5. *Nonoperating Interest*

A nonoperating interest is a term that may also have dual meanings depending on its usage. A nonoperating interest in taxation refers to royalty, overriding royalty, and net profits interests. That is, any interest that is not burdened with the cost to find, develop and produce the minerals. A nonoperator or nonoperating interest may also mean the working interest owners that are not responsible for the day to day operational decisions that the operator of a well must make. Working interest owners, however, are burdened with the cost of finding, developing, and producing the minerals.

7.2. **OPERATING AGREEMENTS**

A clear and comprehensive operating agreement is a necessary ingredient to any oil and gas deal. In most cases, several entities share the costs and revenues associated with an oil and gas lease. In these cases, an operator must be designated to carry out the terms and obligations of the lease for the parties. An operator as discussed earlier, carries on the day to day management activities for the lease. The operator's specific duties, and limitations should be clearly specified. Model Form Operating Agreements such as the A.A.P.L. (American Association of Petroleum Landmen)

Form 610 are available. The following paragraphs will discuss some of the important clauses in an operating agreement.

7.2.1. *Cost And Revenue Sharing*

A table showing the working interest (WI) and net revenue interest (NRI) from each party should be clearly presented in the operating agreement. This will eliminate any confusion that may exist. In this text, the working interest as well as the revenue interest must sum to 100%. That is, every dollar of cost and every dollar of revenue must be accounted for. A systematic technique to determine WI's and NRI's is presented later in the chapter.

7.2.2. *Operator Responsibilities*

The designated operator is required to conduct the operations in a prudent and deligent manner. Often the agreement will specifically outline the operators duties associated with the drilling and development of the initial well as well as subsequent operations. Any drilling done must be on a competitive basis. If the drilling is performed using equipment owned by the operator, the charges need to be competitive.

7.2.3. *Expenditure limits*

The operating agreement limits the amount that the operator can spend on a single project before prior approval is required of all parties. Drilled or deepened wells usually require the consent of all parties subject to certain non-consent clauses that are discussed in the following section.

7.2.4. *Non-Consent Clause*

Any party in an operating agreement may propose to drill or deepen a well in the operating area. If all parties do not consent to the proposed work, a provision of the agreement should specify how the costs and revenues are to be shared if the project is undertaken. Each consenting party is given the option of taking its proportionate share of any non-consenting parties interest in the well. If after this analysis, there is insufficient participation, the proposing party may withdraw its proposal.

The agreement provides that where less than all parties consent, the consenting parties will own the production of the non-consenting parties in an agreed upon amount until certain conditions are met. The nonconsenting parties share of costs and revenues is given to the consenting parties until the net operating income BFIT from the non-consenting parties share of production equals X% of the non-consenting parties share of the cost to drill and complete, and Y% of the non-consenting parties share of the cost of the surface equipment. Often X% is 300 to 500% (a non-consent penalty) and Y% is 100%. After the conditions are met, then ownership reverts to the non-consenting parties as before.

7.2.5. *Accounting Procedures*

The operating agreement will usually refer to the COPAS or accounting procedures exhibit. A COPAS is a set of standards developed by the Council of Petroleum Accountants Societies of North America. The COPAS addresses how direct costs, overhead, and the pricing of material purchases, transfers, and dispositions will be handled.

The agreement should be clear concerning what is to be included in the overhead charge versus what items are to be billed as direct charges. For property transfers, material is classified according to its condition and usability.

7.3 A TECHNIQUE FOR DETERMINING WI AND NRI

First, it is important to note that different terminology is used in this text on oil property evaluation than is commonly used by land and legal departments. The objective of an oil property evaluation is to determine profit (defined as NCF) for each period. To accomplish this, we need to know how costs and revenues are shared for each phase of a deal (before payout and after payout, for example). For this reason the method and terminology used in oil property evaluation establish a working interest and net revenue interest for each player in the deal. A player's working interest (WI) is his fractional share of 100% of the costs, and a player's net revenue interest (NRI) is his fractional share of 100% of the revenues.

FIGURE 7-2

WI and NRI Terminology Used By Land Department

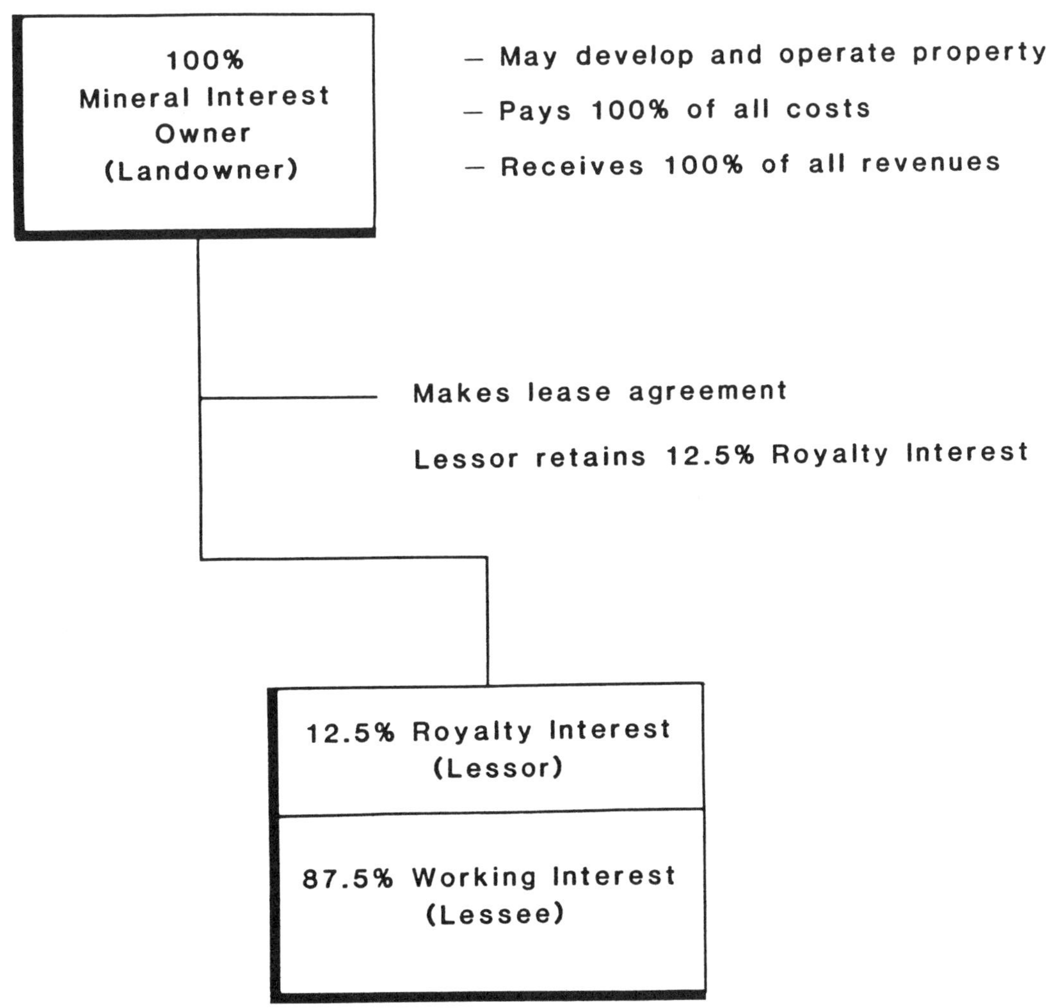

The 87.5% Working Interest Owner pays 100% of all costs **and** collects 87.5% of all revenues.

The 12.5% Royalty Interest Owner pays no costs **and** collects 12.5% of all revenues.

The WI and NRI terminology used by land and legal departments is *not* consistent with this approach. Instead of separating cost and revenue sharing so that the summation of all the WI's equals 100% and the summation of all the NRI's equals 100%, they ignore the cost part of the deals. They separate the revenue from the well into two parts — the royalty and the "working interest." The royalty is that portion of the gross revenues which are paid to the owner(s) of the mineral interest. The "working interest" is that portion of the gross revenues which are paid to people other than the owner(s) of the mineral interest. Thus the sum of the working interest plus the royalty interest equals 1.0 when dealing with land and legal. This can lead to some confusion. Remember that this book refers to working interest as the share of the costs — not the share of gross revenue. The difference in oil property evaluation and land department terminology is demonstrated by Figures 7-1 and 7-2.

Determining your working and net revenue interest in a deal can be quite difficult ("you can't tell the players without a scorecard"). A technique which helps the authors visualize the flow of interests is discussed in the following paragraphs. While it doesn't cover all cases, it does help in most. The technique is three-fold. First a "tree" diagram is drawn to show the flow of the deal. Second, a flowchart is drawn to show the interests of all parties on our branch of the tree. Third, a table is constructed showing the working and net revenue interests of all of the parties.

7.3.1 *The Tree*

An upside down tree (with its root at the top) is drawn to show the relationship of the parties. It always starts with the owner of the mineral rights and proceeds down through the various brokers and dealers until it reaches you. If you pass on some of the deal, it will extend below you. In order to show how the technique works, let's look at a deal.

A, a landowner, leases the right to explore for and produce hydrocarbons to D, retaining a 1/8 royalty interest. D conveys 100% of the lease to E, retaining a 5% (of 8/8) overriding royalty interest. E wants to lay off part of the lease so he teams up with F. E sells F one-fourth of E's interest "straight up". That is, E does not add any additional burdens to the lease. Let's draw the tree for this deal.

Notice that the flow of the deal is quite simple. A more typical deal might look like Figure 7-3.

In the case of Figure 7-3, the tree is quite complicated. However, *we are only interested in that portion between us and the root* and *one level below us*. For the deal shown in Figure 7-3, if we are "E" and we "prune" the tree we get

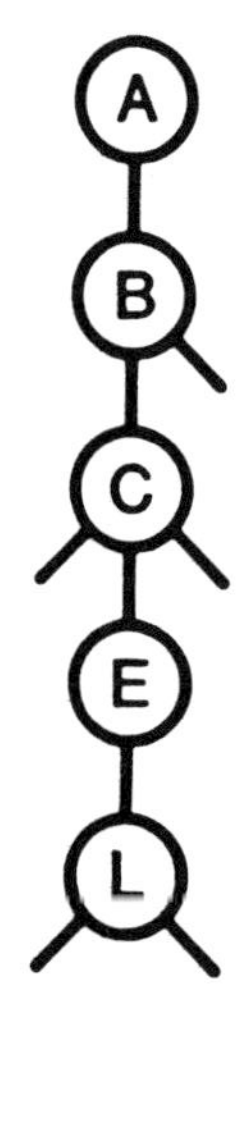

This is the only portion of the tree that E is interested in.

Back to our simple example. We have drawn the tree and identified the branch in which we are interested (trivial in this example).

7.3.2 *The Flowchart*

Now let's draw a flowchart to show how the working and net revenue interests flow through the system. The convention that we will use is as follows:

Cost Side	The Player	Revenue Side
WI rcvd		Last Burden
WI kept	Name	LRI
WI passed on		Burden added by the player

where

WI is working interest

LRI is lease revenue interest (% of 100%)

Note: All values are expressed as percent of 8/8.

These "blocks" are hung from one another just like the portion of the tree in which we are interested. For our simple example the first block is:

a	100		
b	0	Landowner	100
c	100		12.5

On the cost side the landowner "received" 100% of the working interest, passed 100% on and therefore kept 0%. Note that b = a − c or WI kept = WI rcvd − WI passed. On the revenue side the lease was burdened 0% when he got it, so he has a "100% lease". That is, if the landowner had decided to develop the property (rather than create a lease agreement), the landowner would receive 100% of the revenues. The landowner, however, leased the property and he burdens it with a 12.5% royalty interest as it passes him. Now let's add the next block.

FIGURE 7–3

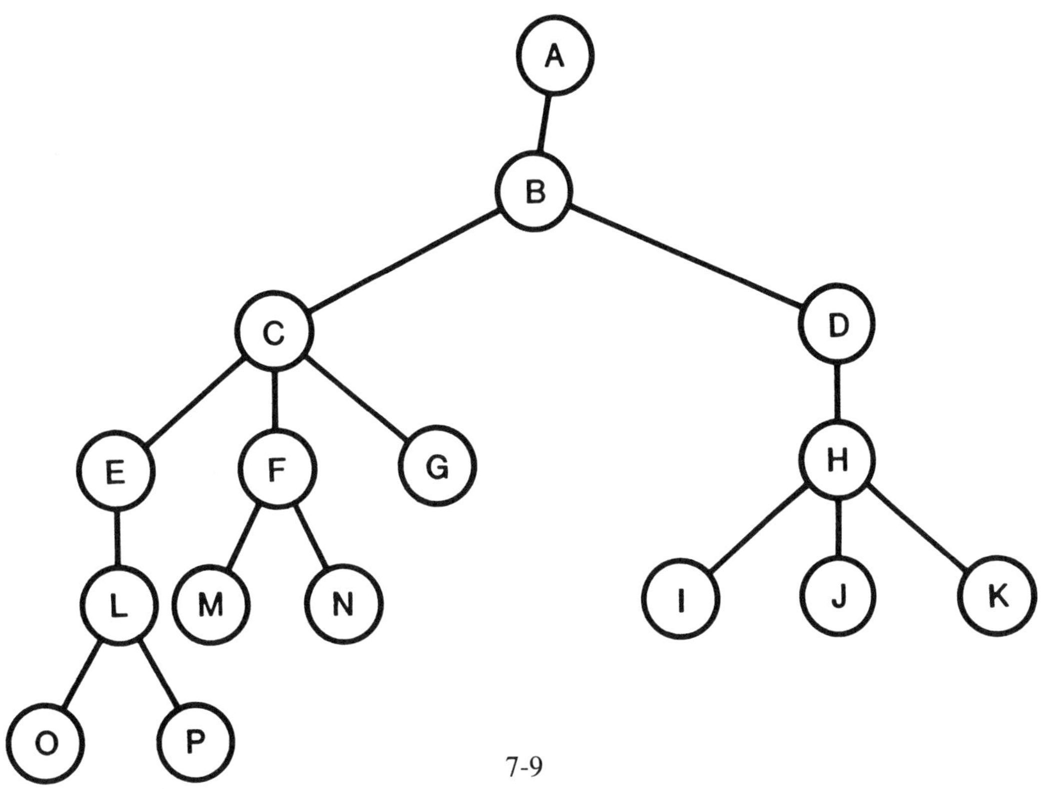

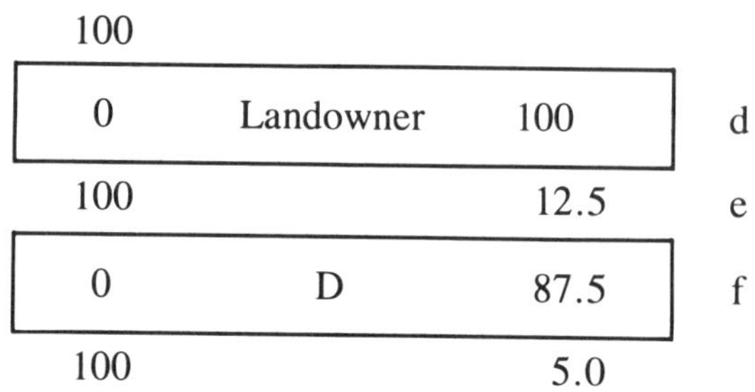

Notice how the blocks mesh together. D passes 100% of the working interest and burdens it with a 5% ORR proportionately reduced. Note on the Revenue side that f = d—e or LRI = Last LRI—Last Burden. Let's continue.

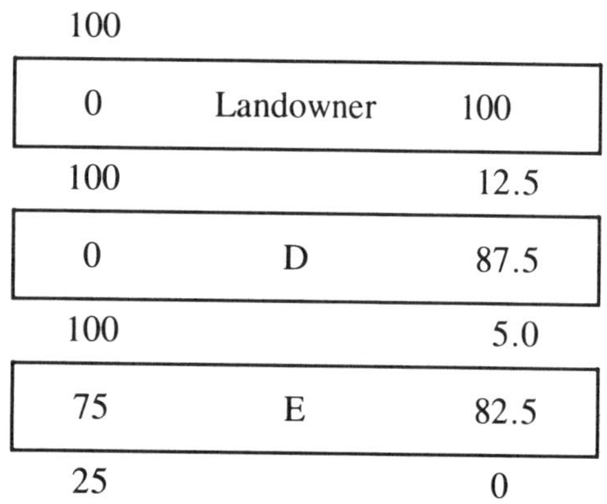

E keeps 75% of the deal and passes 25% on. He does not place any additional burdens on the lease.

With the addition of the last block the flowchart becomes:

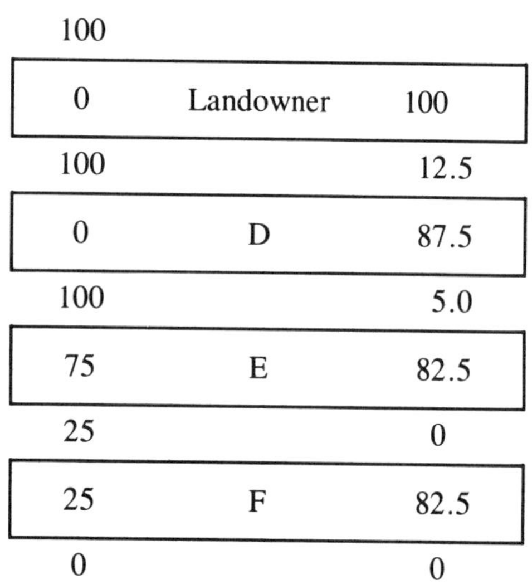

F keeps the last 25% of the deal.

7.3.3 *Table of WI's and NRI's*

We can now construct a table of everyone's working and net revenue interests. The completed table is:

	L.O.	D	E	F
WI	0	0	75	25
NRI	12.5	5.0	61.875	20.625

The working interests are read directly from the boxes of the appropriate player. The net revenue interests for each player can be found by using the following algorithm.

1. Multiply WI kept * LRI (from box)
2. Multiply WI passed * Burden Added (below box)
3. NRI = Results of step 1 plus step 2

In the example:

Landowner:

NRI = WI kept * LRI
+ WI sold * burden added
= 0 * 100 + 100 * 12.5
= 12.5%

Player D:

NRI = 0 * 87.5 + 100 * 5
= 5%

Player E

NRI = 75 * 82.5 + 25 * 0
= 61.875%

Player F

NRI = 25 * 82.5 + 0 * 0
= 20.625%

This technique works for most cases if the rules are followed.

Rule 1: All working interests are expressed as a percent of 100%. (None of this "I'll take 25% of your 1/6 of 1/3 of the deal. Work it out; i.e., "I'll take 1.3888889% of 100% of the deal.")

Rule 2: All burdens are expressed as a percent of 100%, proportionately reduced. (Not "I'll retain a 1/16 of 7/8 overriding royalty." Admit you're sticking them for a 5.46875% of 100% override, proportionately reduced, of course).

7.3.4 *Phases of a Deal*

Farmouts, carried interests and other forms of reversionary interests are very common. Deals such as these make it necessary to talk about how costs and revenues are shared for the various phases of the deal (such as before casing point, after casing point but before payout, and after payout). Casing point and before payout in this case are called trigger points. In deals such as these, we must determine the WI's and NRI's for each phase of a deal. Figure 7-4 shows the time periods and trigger points for an example deal:

Figure 7-4

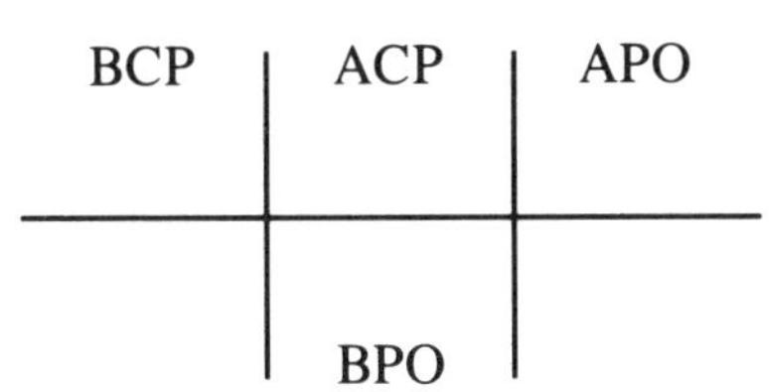

For each player:

WI = ? WI = ? WI = ?

NRI = ? NRI = ? NRI =?

Examples 7-1 and 7-2 demonstrate the concept of a carried interest and an ORRI converted to a working interest and determining WI's and NRI's for the various phases of a deal.

7.3.5 *Complex Deal*

Now let's try again with a somewhat more complex deal. This one includes a farm out and a carried interest.

A lease broker, Z, obtains a lease for a 12.5% landowners royalty. He sells 100% of the deal to A retaining a 2 ½ % ORR. A farms out 50% of the deal to B at 1/3 for 1/4 and 25% of the deal to C at 1/3 for 1/4 retaining a 5% ORR (on both farm outs) until payout. C sells 10% (of 100%) to each of J and K retaining a 2 ½ % ORR. J sells 5% to L heads up and K sells 5% to M retaining a 1% ORR. B sells 5% of the deal to D heads up, 5% to E retaining a 5% ORR, and 30% to Y for a 10% carry to the casing point. You are Y. Y sells 10% of the deal to G retaining a 5% ORR. G sells 2% to H retaining a 5% ORR and 2% to I retaining a 3% ORR. What is your WI and NRI at all times?

Whew! This looks totally impossible. After the panic subsides, the first step is to plot the tree as shown on Figure 7-5 (next page).

Notice that the branch you are on has been highlighted. You (Y) only care about this branch. Your deal is independent of C,J,L,K,M,D,E,H, and I. Get rid of them! You are left with the following

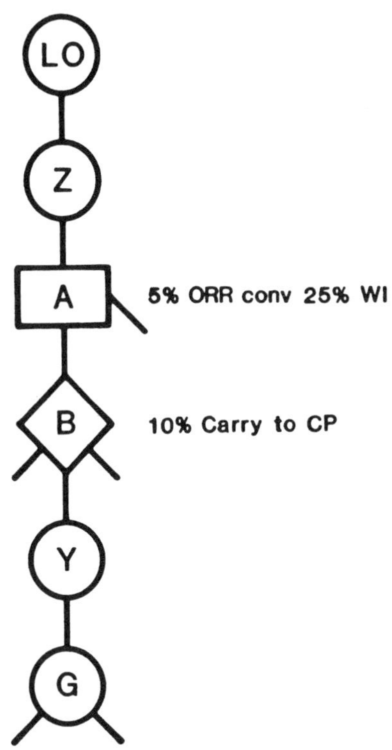

FIGURE 7–5

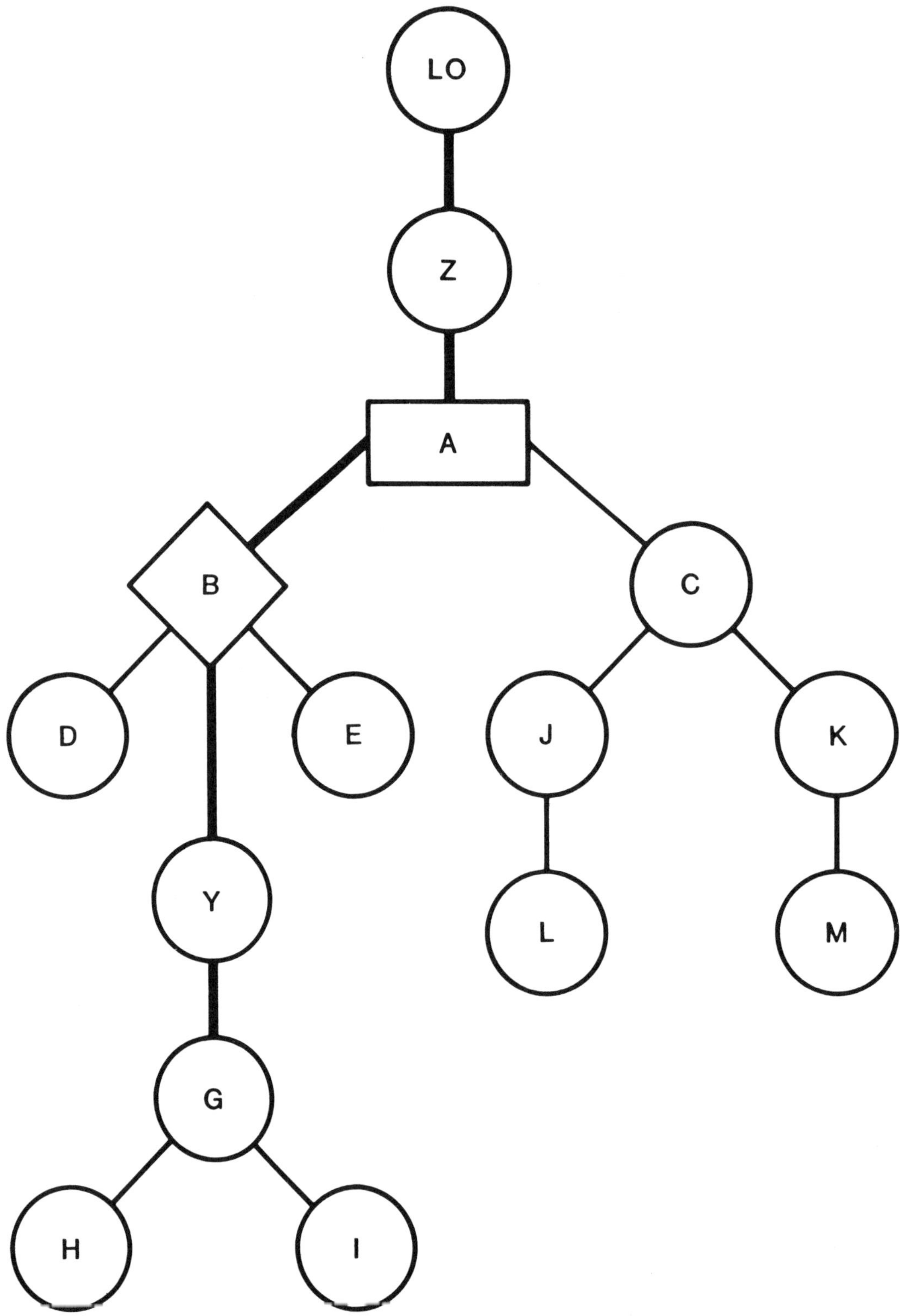

Example 7-1
Carried Interest to Casing Point

A piece of property in which the landowner (LO) has the 100% mineral interest is subject to the following deal.

- Landowner retains 12.5% royalty interest.
- Broker (B) sells the entire deal (100%) to Major Oil Company (MOC) retaining a 5% ORRI proportionately reduced.
- Major Oil farms out 100% of the deal for a 25% carried interest, proportionately reduced, to the casing point to Independent Oil Company (IOC).

Prepare the flowchart and table the WI and NRI interest for this deal for the following periods:

1) Period before the casing point, and

2) Period after the casing point.

1. **Period Before Casing Point**

	Cost Side	Player	Revenue Side	
	100			
	0	LO	100	
WI Rec'd	100		12.5	
WI Retained	0	B	87.5	LRI
WI Passed	100		5.0	Burden Added
	0	MOC	82.5	
	100		0	
	100	IOC	82.5	

Player	WI (from Box)
Landowner	0.000
Broker	0.000
Major Oil	0.000
Independent Oil	1.000

Example 7-1
(Continued)

2. **Period After Casing Point**

	Cost Side	Player	Revenue Side
	100		
	0	LO	100
	100		12.5
	0	B	87.5
WI Rec'd	100		5.0
WI Retained	25	MOC	82.5
WI Passed	75		0
	75	IOC	82.5

Player	WI (From Box)	NRI	Box (WI*LRI	+	Below Box WI*Burden Added)
Landowner	0.000	0.12500	(0*1.00	+	1.00*.125)
Broker	0.000	0.05000	(0*.875	+	1.00* .05)
Major Oil	0.250	0.20625	(.25*.825	+	.75*0)
Independent Oil	0.750	0.61875	(.75*.825	+	0*0)
	1.000	1.00000			

Major Oil's 25% carried interest to the casing point means that after the casing point, it will pass only 75% of what it passed for the period before the casing point. In our case, 0.75 of 1.0 = .75. Major Oil therefore retains 25% after the casing point.

Now it looks a little more manageable. Let's draw the flowchart.

Because of the terms of the deal to you, there are three separate periods which must be flowcharted. These are: 1) before the casing point; 2) from the casing point until payout; 3) after payout. Period 1 arises because of the deal between B and Y (carried interest to casing point). Periods 2 and 3 arise from the deal between A and B (a back in after payout).

Let's flowchart the deal before the casing point. The net revenue interest is immaterial at this time since no revenues have been generated, so we only worry about WI. As discussed earlier, a 10% carry means the working interest of the carrying party gets decreased by 10% at the time some predefined condition or trigger point (the casing point) is reached. If we are trying to find *only your interest*, the flowchart would look like:

Period Before Casing Point

	Cost Side	Player	Revenue Side
rcvd	100		
kept	0	LO	
sold	100		
	0	Z	
	100		
	50	A	
	50		
	20	B	
	30		
	20	Y	
	10		
		G	

So you have to pay 20% of the dry hole costs. Now let's draw the diagram for the period after the casing point and before payout occurs; again this diagram can be used only to find *your interest*. (We'll fix that later.)

After Casing Point & Before Payout

	Cost Side	Player	Revenue Side
rcvd	100		
kept	0	LO	100
sold	100		12.5
	0	Z	87.5
	100		2.5
	50	A	85.0
	50		5.0
	23	B	80.0
	27		0
	18	Y	80.0
	9		5.0
		G	

Let's look at the Cost side. The easiest way to get your WI is to subtract 10% of your original interest. (20% − 10% of 20% = 18%.) However, if we work the diagram assuming only your branch exists, then, post-casing point, it appears that B only sold 27% to you and you sold 1/3 of that or 9% to G.

On the Revenue side you can see all of the overrides which have been attached to the lease. To find your net revenue interest multiply the WI in your box by the LRI in your box and add the product of the WI and the burden added below your box.

$$\begin{aligned} NRI &= 18\% * 80 + 9\% * 5 \\ &= 14.85\% \end{aligned}$$

So after the casing point and before payout, your WI is 18% and your NRI is 14.85%.

After payout the 5% ORR in the deal between A and B converts to a 25% WI. That means your WI decreases by 25% (to become 13.5%). Let's draw the flowchart decreasing the WI transferred from A to B, B to Y, and Y to G by 25%.

After Payout

Cost Side	Player	Revenue Side
100		
0	LO	100
100		12.5
0	Z	87.5
100		2.5
62.5	A	85.0
37.5		0
17.5	B	85.0
20.25		0

13.50	Y	85.0
6.75		5.0
	G	

Notice that there is now no override between A and B since it was converted to a 25% WI. Our working interest is 13.50% and our net revenue interest is 11.8125% (0.135 * 0.85 + 0.0675 * 0.05).

Now let's complete a table for our interests.

Y's Interests

	WI	NRI
Before Casing Point	20	—
Casing Point to Payout	18	14.85
After Payout	13.50	11.8125

Example 7-2
5% ORRI Converted to 25% WI on Payout

A landowner (LO) leases the right to explore for and produce hydrocarbon to Broker (B) for a 1/8 royalty interest. Broker B sells 100% of the deal to Major Oil Company (MOC) for a 5% ORRI. Major Oil farms out all of the deal for a 5% ORRI convertible to a 25% WI on payout, proportionately reduced, to Independent Oil Company (IOC).

Prepare the flowchart and table the WI and NRI's for the following periods:

1. Period Before Payout, and
2. Period After Payout.

1. **Period Before Payout**

Cost Side	Player	Revenue Side
100		
0	LO	100
100		12.5
0	B	87.5
100		5.0
0	MOC	82.5
100		5.0
100	IOC	77.5

Player	WI (From Box)	NRI	Box (WI*LRI	+	Below Box WI*Burden Added)
Landowner	0.00	0.125	(0*1.00	+	1.0*.125)
Broker	0.00	0.050	(0*0.875	+	1.0*.05)
Major Oil	0.00	0.050	(0*0.825	+	1.0*.05)
Independent	1.00	0.775	(1.0*.775	+	0*0)
	1.00	1.000			

So we put up 20% of the risk money to receive 11.8125% of the gross income after payout. This is not too stiff if the 1/3 for 1/4 and 10% carry apply only to the first well. About the best you can do is put up 20% of the risk money and receive 13.125% of the gross income after payout. (That's a 25% back-in on a 87.5% lease.)

If you are evaluating deals notice how much easier it is to use the table shown above rather than work your way through the deal. Insist that people showing you deals provide the above table. If nothing else, you'll get the satisfaction of seeing them sweat their way through the deal.

The previous flowcharts are only valid for your interest. That is because we did not include any of the other branches on the tree. Let's draw the full flowcharts for our example. (Figures 7-6, 7-7, and 7-8)

Notice that the expanded version of the flowchart looks like a tree. We can now find everyone's working interest by reading the number in the box and everyone's net revenue interest by multiplying their working interest by the lease net revenue interest and adding the product of the working interest passed on times the additional burden added. Let's look at Table 7-1 for this deal.

Example 7-2
(Continued)

2. **Period After Payout**

Cost Side	Player	Revenue Side	
100			
0	LO	100	
100		12.5	
0	B	87.5	
100		5.0	
25	MOC	82.5	
75		0	← Note
75	IOC	82.5	

Player	WI (From Box)	NRI	Box (WI*LRI	+	Below Box WI*Burden Added)
Landowner	0.00	0.12500	(0*1.00	+	1.0*.125)
Broker	0.00	0.05000	(0*1.00	+	1.0*.05)
Major Oil	0.25	0.20625	(.25*.825	+	.75*0)
Independent	.75	0.61875	(.75*.825	+	0*0)
	1.00	1.00000			

Note that the 5% ORRI burden added by MOC disappears when the conversion to a working interest is made. The 5% ORRI convertible to a 25% WI is also called a 25% back-in. That is, after payout MOC will pass 75% of what it had previously passed or take back 25% of what it had passed.

Table 7-1
Summary of WI's and NRI's for Complex Deal

	Before Casing Point	Before Payout		After Payout	
	WI	WI	NRI	WI	NRI
L.O.	0	0	12.500	0.00	12.50000
A	25	25.0	25.000	43.75	37.18750
B	10	13.0	10.650	9.75	8.47500
C	5	5.0	4.500	3.75	3.56250
D	5	5.0	4.000	3.75	3.18750
E	5	5.0	3.750	3.75	3.00000
G	6	5.4	4.194	4.05	3.34800
H	2	1.8	1.260	1.35	1.01250
I	2	1.8	1.296	1.35	1.03950
J	5	5.0	3.875	3.75	3.09375
K	5	5.0	3.925	3.75	3.13125
L	5	5.0	3.875	3.75	3.09375
M	5	5.0	3.825	3.75	3.05625
Y	20	18.0	14.850	13.50	11.81250
Z	0	0.0	2.500	0.00	2.50000
Tot.	100	100.0	100.000	100.00	100.00000

Notice that the sums of the interests are all 1.0. This has to be. If they don't add up to 1 then something has not been calculated properly.

The method shown above helps show the organization of the deal and helps determine the WI and NRI for all the players. It works well for normal deals but it is not foolproof. As more practice in working out deals is obtained it will become easier and easier.

FIGURE 7–6

Before Casing Point

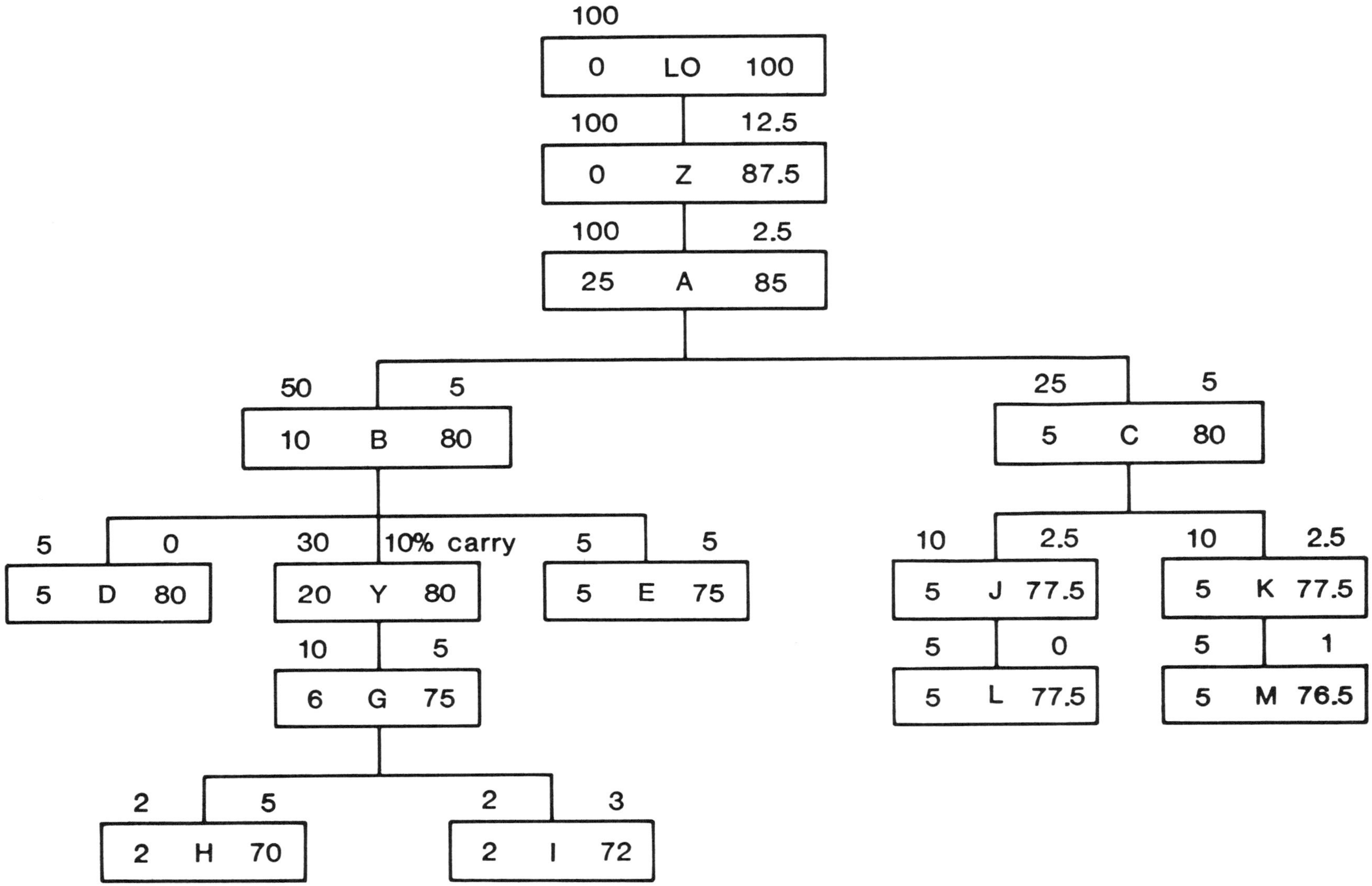

FIGURE 7-7

After Casing Point, Before P.O.

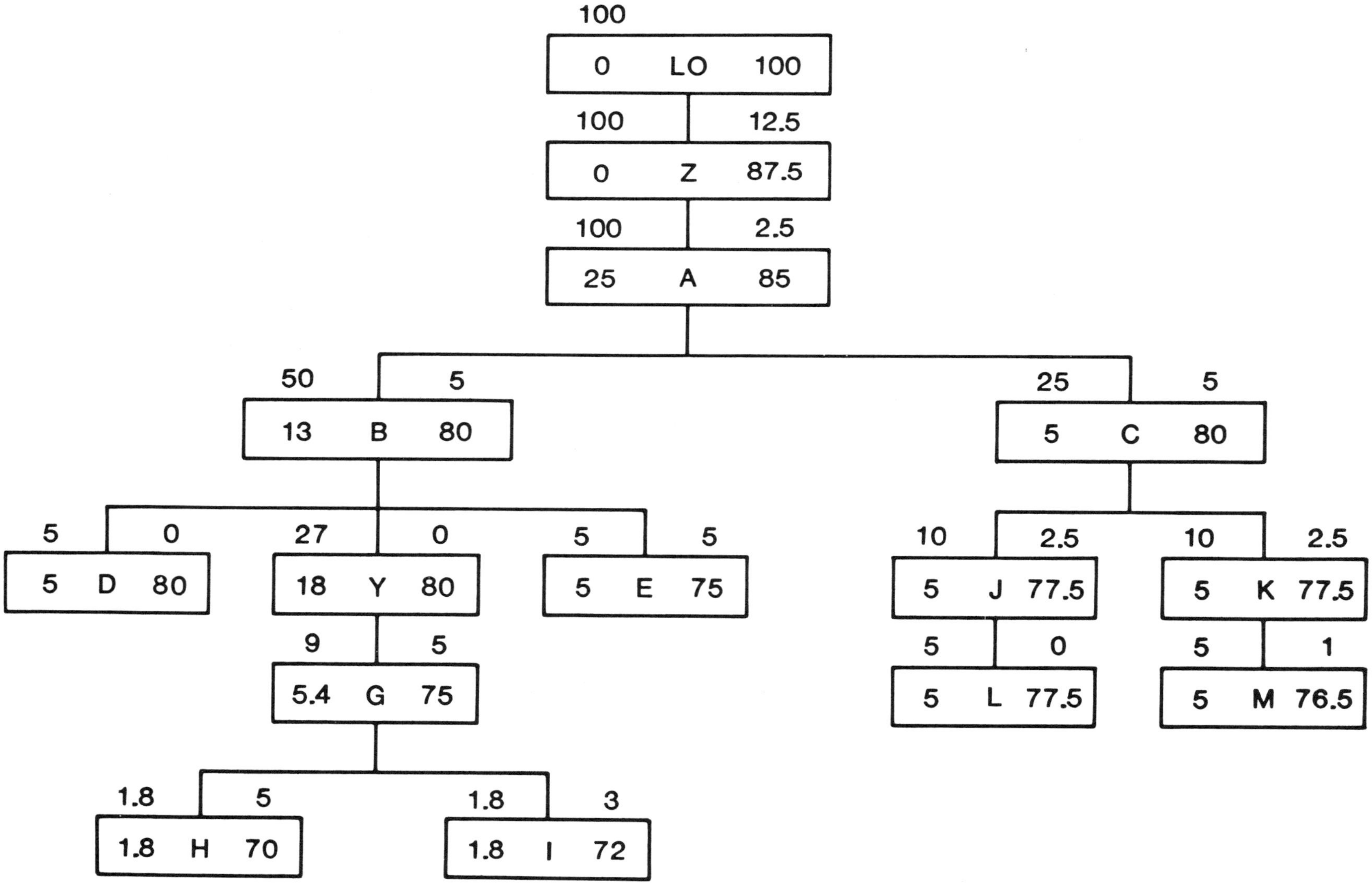

FIGURE 7–8

After P.O.

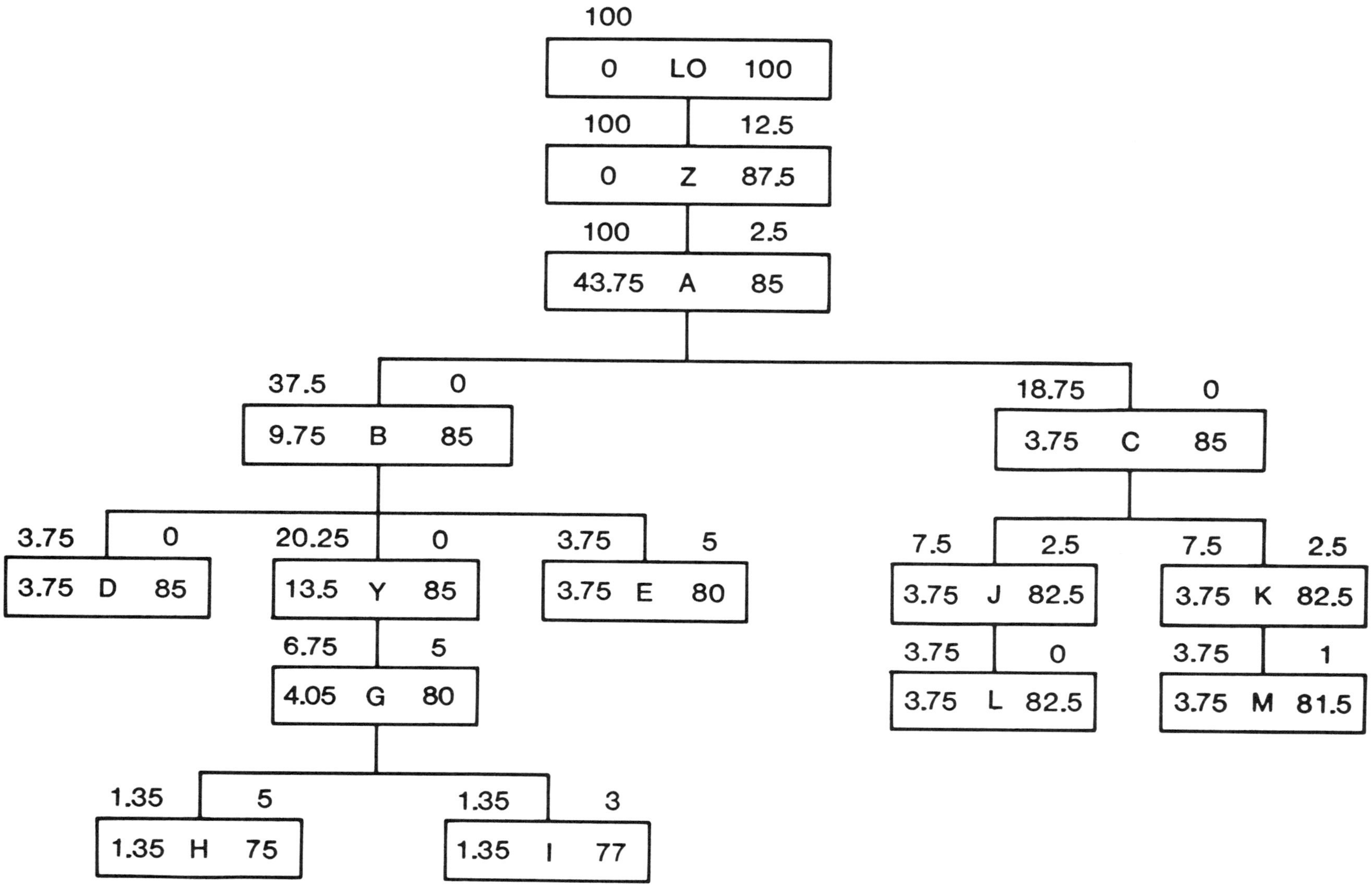

Problems

7-1 A broker obtains a lease from a landowner who retains a 12½% royalty interest. The broker turns 100% of the deal to Major Oil for a 5% overriding royalty interest proportionately reduced. Major Oil wants to reduce its risk so it farms out its interest for a 10% carried interest to the casing point to Independent Oil, the farmee.

a. Determine what fraction of the costs each player must pay prior to setting casing.

b. Determine the working interest and revenue interest for each player from the casing point forward.

7-2 A broker obtains a lease from a landowner who retains a 12½% royalty interest. The broker turns 100% of the deal for a 2½% overriding royalty interest proportionately reduced to an independent. The independent farms the deal out to Farmee No. 1 for a 5% override convertible to a 25% WI on payout proportionately reduced.

a. Determine the working and revenue interest for each player until payout occurs. How is payout calculated for the reversion?

b. Determine the working and revenue interest for each player after payout occurs.

c. Assume that we add two more players. Farmee No. 1 makes a deal with Farmee No. 2. Farmee No. 2 will take 80% of the deal and carry Farmee No.1 for 10%, proportionately reduced, to the casing point. Farmee No. 2 then gives City Slicker 22% of the deal for a 10% override proportionately reduced. For each stage of the deal show how the costs and revenues will be shared by each of the parties.

d. Assume the "dry hole" cost is $200,000, the cost to complete is $300,000, (total cost is $500,000) the oil price is $30/bbl, and operating costs are $3 per barrel. Determine, (1) each players share of the dry hole cost, (2) the sharing of costs and revenues to payout, and (3) how much production the well must have before City Slicker breaks even.

e. At the reversion point for the deal between the Independent and Farmee #1 ("payout"):

How much investment has each player made?

How much income (revenue less operating costs) has each player received?

7-3 You are a 100% working interest owner of a lease. However, you are not sure if you want to take a 100% working interest (you pay all the costs) and an 87.5% net revenue interest (you receive 87.5% of the revenues) or farmout and take a carried interest until payout. Under the carried interest option you would receive a 5% overriding royalty (receive 5% of the revenues) until payout occurs. Payout is based on a 100% working interest owner who has an 82-1/2% net revenue interest. On payout your royalty interest would change (revert) into a 25% WI and a 21.875% net revenue interest. Use the production schedule provided. The well costs $350,000 to drill and complete. Assume annual cash flows and end of period discounting. Operating costs are $3/bbl (Includes state and local taxes and the windfall profits tax). Oil price is $27/bbl.

a. Which deal do you prefer if your reinvestment opportunity rate is 15%. Do your analysis BFIT.

PRODUCTION SCHEDULE

Time (Years)	Gross Production (bbl)
1	16,251
2	5,030
3	3,599
4	2,847
5	1,679
6	500

b. Plot the NPV discounted at 15% for each case versus reserves. For scheduling, if reserves double, assume that production doubles in *each* year using the above production schedule. Repeat for various reserve assumptions. At what level of reserves are you indifferent between keeping 100% of the deal and the carried interest. Solve graphically.

Table of Contents
Chapter 8
Costs

8.1 Investments 8-1
8.2 Operating Costs 8-1
8.2.1 Economic Limit 8-2

Chapter 8

Costs

The cash flow model for an oil property evaluation has certain costs which are attributable to your working interest. The objective of this chapter will be to discuss two of these costs, operating costs and investments, and their treatment in the evaluation process.

8.1. INVESTMENTS

Capital expenditures should be classified as investments. An investment is an expenditure that will generate benefits for more than one period (usually one year). The cost of a machine which has a useful life of 5 years will help generate revenue (hopefully) over the next five years and would therefore be classified as a capital expenditure. Salaries, on the other hand, are deemed to benefit only the current period and are classified as an expense rather than a capital expenditure. When classifying an expenditure as an expense or a capital item, ignore the income tax treatment of the item. The tax treatment will be presented in Chapter 11. The income tax treatment is a separate problem with its own set of rules. The tax rules are used only to generate the amount of cash income taxes that must be paid during the taxable year.

The following is a list of some capital expenditures found in oil property evaluations.

Cost of drilling and developing the well:
- drilling mud
- casing
- cement
- well logging
- completion costs
- tubing
- wellhead

Cost of surface equipment:
- tanks
- separators
- compression equipment
- transfer pumps
- artificial lift equipment
- metering equipment
- buildings
- flowlines
- labor to install surface equipment

Often the engineer is requested to prepare an estimate to drill and complete a well. This is usually done in the form of an authority for expenditure (AFE).

A sample AFE is shown as Exhibit 8-1. The estimates for AFE's are usually obtained from recent expenditures in the same area where the drilling will occur, drilling contractors, service companies, and current equipment catalogs.

8.2. OPERATING COSTS

Operating costs are those expenditures that will only benefit the period in which they are made. Two typical cost behavior patterns are variable costs and fixed costs. Variable costs increase with an increase in activity. Raw material used in manufacturing a commodity is an example of a variable cost. The more units produced the greater the total cost for raw material. A fixed cost, however, remains constant for a range of activity. Management and the accounting staff are two examples of fixed operating costs. The addition of a field may not require any change in the size of management or the accounting staff. There may, however, be a level of activity such that the accounting staff will be increased in size. Thus fixed costs are usually only fixed for a range of activity.

Some examples of operating costs are:

- Direct labor for personnel responsible for the lease.
- Maintenance, which includes costs for cleaning tanks, engine maintenance, and painting.
- Supplies, such as filters, oil for oil changes, and charts for meters.
- Workovers and recompletion costs.
- Handling or treating produced water and utility costs.
- Fixed charges per well month for operating agreements.

The treatment of some of these costs in the property evaluation, including the decision to abandon, deserve some comments. Every evaluation, as emphasized in earlier chapters, is an incremental analysis. The decision to buy should be made by looking at the after tax net cash flow for the company with the project, versus the after tax net cash flow for the company without the project. This incremental approach will result in a determination of the relevant costs. That is, when a proposed purchase is evaluated, what costs will increase, decrease, or go away as a result of the acquisition. A look at the accountant's lease operating expense statements may show an amount for allocated overhead. Allocated overhead includes the salaries for management, staff engineers, and staff accountants. If any of the amounts included in overhead will increase as a result of the acquisition, then the project's cash flow must be burdened with these incremental costs.

One final comment concerning operating costs. Operating costs should be treated on a total cost basis rather than on a per unit (barrel or MCF) basis. Unit costs can be misleading. It will cost about the same whether or not a well produces 80 barrels per day or 10 barrels a day. If the expected total cost of $2,000 per month is put on a per unit basis when the well produces 80 barrels per day, the operating cost would be approximately $0.82 per barrel. When the production declines to 10 barrels per day, the total monthly operating cost calculated at $0.82 per barrel of oil would only be $250. In reality the total monthly operating costs might easily increase above $2000/month.

8.2.1. *Economic Limit*

The decision to abandon a well should also be based on "relevant" costs and revenues. That is, only revenues and costs that will go away if a well is abandoned should be considered. Once again an incremental analysis will determine the relevant revenues and costs. For example, if a well on a lease is to be abandoned, should the pumper's salary now allocated to the well be considered in the analysis? To answer the question do an incremental analysis. What are the revenues and costs for the entire lease with the well less the revenues and costs for the lease without the well. If the pumper will be retained whether or not we abandon the well, then the allocated salary is not a relevant cost. The following equation can be used to calculate the producing rate at which the revenues exactly equal the costs. This producing rate is called the economic limit.

Your share of revenues = your share of costs.

$$(Q_{el})\ (1\text{-}PTR)\ (SP)\ (NRI) = \frac{OPC}{30.4}\ (WI)$$

Rearranging:

$$Q_{el} = \frac{(OPC)\ (WI)}{(30.4)\ (1\ \text{-}PTR)\ (SP)\ (NRI)}$$

Where:

Q_{el} = economic limit (bbl/day)
SP = sales price ($/bbl)
NRI = your net revenue interest
OPC = monthly operating costs ($/mo.)
WI = your working interest
PTR = production tax rate (fraction of sales price)

Exhibit 8-1
AUTHORITY FOR EXPENDITURE

Date________________

AFE No.__________ Lease Name______________________________ Well No.________

Description______________________________ Parish/County______________

State______________ Area______________________ Operator______________

Project______________________________ Prepared by______________

Exploration Development Results: Oil Gas Dry T.D.________ Date________

DRILLING INTANGIBLES:	Dry Hole Without Pipe	Completed Well	Actual Cost	Over (Under)
Administrative Overhead				
Rig Move				
Location-Stake/Permit/Damages/Legal				
Location-Const/Dirtwork/Road/Marker				
Turnkey______to______ft				
Footage______ft @ $______/ft				
Day Work______days WDP @ $______/day				
______days WODP @ $______/day				
Company Labor & Equipment				
Contract Labor & Equipment				
Cement & Cementing				
Testing Service & Equipment				
Coring Services & Equipment				
Logging				
Geologist/Geological Services				
Engineer/Engineering Services				
Equipment Rental				
Well Control Equipment				
Mud, Water, Fuel				
Casing______ft				
______ft				
______ft				
Welding & Casing Crew				
Casing Accessories				
Bits #-______ Size-______				
#-______ Size-______				
#-______ Size-______				
Transportation				
Miscellaneous				
Total Drilling Intangibles				

Exhibit 8-1
(Continued)

AFE NO. ______	Dry Hole Without Pipe	Completed Well	Actual Cost	Over (Under)
COMPLETION INTANGIBLES:				
Administrative Overhead				
Rig Move				
Permanent Surface Damages				
Completion Unit ______ days @ $ ______				
Company Labor & Equipment				
Contract Labor & Equipment				
Cement & Cementing				
Testing Services & Equipment				
Logging, Perforating, Wireline				
Fracturing/Acidizing				
Engineer/Engineering Services				
Equipment Rental				
Fuel, Water, Power				
Welding & Casing Crew				
Battery Construction & Dirt Work				
Transportation				
Miscellaneous				
Total Completion Intangibles				
Total Intangibles				

Exhibit 8-1
(Continued)

AFE NO. ____	Dry Hole Without Pipe	Completed Well	Actual Cost	Over (Under)
EQUIPMENT:				
Casing ____ ft				
____ ft				
____ ft				
____ ft				
Tubing ____ ft				
____ ft				
Wellhead Equipment				
Float Equipment, Centralizers, Scratchers				
Packers, Anchors, Seating Nipples				
Rods ____ ft				
____ ft				
____ ft				
____ ft				
Subsurface Equipment - BH Pump, etc.				
Line Pipe ____ ft				
____ ft				
____ ft				
Tanks & Fittings/Stairs/Walks				
Treater & Fittings				
Separator & Fittings				
Gas Processing Unit				
Pumping Unit & Engine				
Other Surface Equipment-Meters, Bldgs, etc.				
Other Valves & Fittings				
Miscellaneous				
Total Equipment Cost				
Total Well Cost				

Table of Contents
Chapter 9
Severance and Advalorem Taxes

9.1 Table—Severance Tax Rates 9-2
9.2 Table—Approximate Advalorem Tax Rates 9-6

Chapter 9

Severance and Advalorem Taxes

State and local taxes levied on the production of oil and gas are referred to as severance taxes, advalorem taxes, conservation taxes, and school taxes among many. The production tax or state and local tax, is deductible from gross income in calculating the net operating income before federal income tax. These taxes are paid regardless of whether or not a profit is made. Accordingly, these amounts should be included as ''relevant'' costs when calculating the economic limit. Tables 9-1 and 9-2 present a summary of the severance and advalorem taxes respectively. Since taxation is a dynamic subject, the appropriate state and Internal Revenue Office should be checked for the specific details. Furthermore, advalorem taxes are hard to quantify since they are often based on assessed values of equipment and property and are administered on a county basis. The tabled values for advalorem taxes are from experience and accordingly should only be taken to be approximations. Many states have complicated exemptions, limits, or sliding scales. In order to provide a quick reference, we have presented a simplified table. API circular TC-10 contains a 26 page summary of severance tax rates. Individual state's departments of revenue should be contacted for details.

TABLE 9-1
Severance Tax Rates by States

Unless otherwise noted rates are stated as fraction of gross value.

State	Oil	Gas	Comments
Alabama	0.06*	0.06	Note 1
	0.10*		Note 2
Alaska	0.1225*	.10	Note 3
	0.15*		Note 4
Arizona	0.025	0.025	
Arkansas	0.04*	0.3¢/MCF	Note 5
	0.05*		Note 6
	5 mill/barrel**		Note 7
California	varies**	Varies	Note 8
Colorado	0.02**	0.02	Note 9
	0.03**	0.03	Note 10
	0.04**	0.04	Note 11
	0.05**	0.05	Note 12
	0.0006*		Note 13
	0.0009*		Note 14
Florida	0.05	0.05	Note 15
	0.08	0.05	Note 16
Georgia	$0.005/bbl.	$.0005/Mcf	
Idaho	0.02*	0.02	
Indiana	0.01*	0.01*	
Kansas	0.08	0.08	Note 39
Kentucky	0.045	0.045	
Louisiana	0.03125	7¢/Mcf	Note 17
	0.0625		Note 18
	0.125		Note 19
	Certain exemption		Note 40
Maryland	None	None	

TABLE 9-1
(Continued)

State	Oil	Gas	Comments
Michigan	0.066	0.05	Note 20
	0.04		Note 21
Mississippi	0.06	0.06	
Montana	0.05	0.0265	Note 22
	0.06		Note 23
	0.0002		Note 24
Nebraska	0.02*	0.02	Note 25
	0.03*	0.03	Note 26
	0.0004	0.0005	Note 27
Nevada	$.005/bbl.**	$.0001/Mcf	Note 28
New Mexico	0.0375	15.2¢/Mcf	Note 29
	0.0019		Note 30
	0.0255		Note 31
New York	None	None	
North Carolina	$.005/bbl.	$.0005/Mcf	Note 32
North Dakota	0.05*	0.05	Note 33
	0.115*		Note 34
Ohio	10¢/barrel**	3¢/Mcf	
Oklahoma	0.07085	0.07085	
Oregon	0.06	0.06	
South Dakota	0.045*	0.045	
Tennessee	0.03*	0.03	
Texas	0.046	0.075	
	0.001875/bbl.**		Note 41
Utah	0.04	0.04	Note 35
Virginia	None	None	
West Virginia	0.0434**	0.0863	Note 36
Wyoming	0.06	0.06	Note 37
	0.04		Note 38
	0.0001		Note 42

Notes for Severance Tax Table 9-1

* Qualifies for windfall profit tax severance tax adjustment.
** Does not qualify for windfall profit tax severance tax adjustment.

NOTES:

1. For wells producing less than 25 barrels per day or 200 MCF per day in unitized fields and all wells that commence production after September 1, 1979. Baldwin County has an additional 1% severance tax.

2. For wells producing more than 25 barrels of oil per day or 200 MCF per day which commenced production on or before September 1, 1979. Baldwin County has an additional 1% severance tax.

3. For oil wells which commenced production after June 30, 1981. For first five years only. Oil gravity and economic limit affect rate.

4. For oil wells commencing production on or before June 30, 1981. Oil gravity and economic limit affect rate.

5. Oil production is 10 BOPD or less during any calendar month.

6. Production is greater than 10 BOPD during any calendar month.

7. The mill tax per barrel does not qualify for the windfall profit tax severance tax adjustment.

8. The severance tax varies depending on needs of Oil and Gas Division. Usually small when compared to the estimated advalorem tax rate. 1983 rate = $0.0144/bbl.

9. For gross income under $25,000. Wells averaging less than 10 BOPD for taxable year are exempt.

10. For gross income > $25,000 and < $100,000. Wells averaging less than 10 BOPD for taxable year are exempt.

11. For gross income > $100,000 and < $300,000. Wells averaging less than 10 BOPD for taxable year are exempt.

12. For gross income $300,000 and over. Stripper production exempt.

13. A conservation tax. For production on or after January 1, 1982.

14. A conservation tax. For production January 1, 1979 and before July 1, 1981.

15. For wells producing less than 100 barrels per day and all oil produced by tertiary methods.

16. For wells producing more than 100 BOPD and not utilizing tertiary methods.

17. For 10 BOPD or less.

18. For 25 BOPD or less and at least 50% salt water.

19. For wells producing more than 25 BOPD and any condensate production.

20. All production not stripper.

21. Stripper production. (Check Michigan rules for specific definition for stripper production)

22. After April 1, 1985

23. Until March 31, 1985.

24. Conservation tax. Montana also has a Resource Indemnity Trust tax of $25 + 0.5% of gross value over $5,000.

25. Wells producing 10 BOPD or less.

26. Wells producing more than 10 BOPD.

27. Conservation tax.

28. A conservation tax.

29. Severance tax. Gas rate changes yearly.

30. Conservation tax.

31. Emergency school tax.

32. A conservation tax.

33. Production tax on all crude oil and gas.

34. Extraction tax of 6.5% on all crude oil except stripper production and royalty interests on first 100 barrels of oil per day in addition to 5% production tax.

35. Occupation tax effective 1/1/84. Wells producing less then 20 BOPD are exempt. Production for the first six months is also exempt. The state has a conservation tax of 0.002.

36. The 0.0434 tax is the business and occupation tax. The gas tax is on the value over $5,000/year.

37. For production where average daily rate is in excess of 10 BOPD for preceding calendar year.

38. For production where average daily rate is less than or equal to 10 BOPD for preceding calendar year.

39. Any advalorem taxes paid may be used as a credit against the Kansas severance tax up to 3.67% of the gross value. Thus the effective severance tax rate may only be 4.33% if the advalorem taxes paid are at least 3.67% of the gross value.

40. A working interest in newly discovered fields are exempt from 50% of the severance tax for 24 months on the first 100 BPD. The 24 month period starts when regular production begins.

41. Oil Conservation tax.

42. Oil and Gas Conservation Tax. Rate varies. Maximum rate is 0.0004.

Table 9-2
Approximate Advalorem Tax Rates
Fraction of Gross Value

State	Oil	Gas
Alabama	None	None
Alaska	0.002	0.002
Arizona	0.05	0.05
Arkansas	0.04	0.04
California	0.09	0.09
Colorado	0.03	0.03
Florida	0.02	0.02
Georgia	0.05	0.05
Idaho	0.05	0.05
Illinois	—	0.08
Indiana	0.07	0.07
Kansas	0.08	0.08
Kentucky	0.06	0.06
Louisiana	None	None
Maryland	0.07	0.07
Michigan	0.02	0.02
Minnesota	0.05	0.05
Mississippi	None	None
Montana	0.09	0.09
Nebraska	0.06	0.06
Nevada	0.06	0.06
New Mexico	0.05	0.05
New York	0.10	0.10
North Carolina	0.05	0.05
North Dakota	None	None
Ohio	0.07	0.07
Oklahoma	None	None
Oregon	None	None
Pennsylvania	0.08	0.08
South Carolina	0.05	0.05
South Dakota	0.02	0.02
Tennessee	None	None
Texas	0.04	0.04
Utah	0.05	0.05
Virginia	None	None
West Virginia	None	None
Wyoming	0.07	0.07

References

Crude Oil Pricing/Windfall Profit Tax Information Service, Federal Programs Advisory Service, Thompson Publishing Group, Washington, D.C.

API, "TC-10 State and Local Oil and Gas Severance and Production Taxes." American Petroleum Institute, 1220 L Street, Northwest, Washington, D.C. 20005.

Table of Contents
Chapter 10
Windfall Profit Tax

10.1 Calculation of Tax 10-1
10.1.1 Removal Price 10-2
10.1.2 Base Price and Adjusted Base Price 10-2
10.1.3 Severance Tax Adjustment 10-3
10.1.4 Net Income Limitation 10-4
10.1.5 Tax Rates 10-8
10.2 Exempt Oil and Transfer Rules 10-11
10.2.1 Exempt Oil 10-11
10.2.2 Transfer Rules 10-11

Chapter 10

Windfall Profit Tax

The Crude Oil Windfall Profit Tax of 1980 (WPTA) was passed by Congress as a part of the oil price decontrol package. The windfall profit tax (WPT) is a temporary excise tax applied to all domestic production of crude oil on or after March 1, 1980 unless specifically exempted. Crude oil not only includes liquids in the reservoir that remain a liquid at surface operating conditions, but also includes condensate liquids, produced from gas wells, that are recovered at any point before a gas processing plant. The tax was imposed to tax any gain realized from the decontrol of oil and also had the impact of reducing the effective price for oil that was already being sold at market price at the time the tax was imposed. The tax will be phased out starting on December 31, 1987 if $227.3 billion has been accrued, or the point in time after this date when $227.3 billion has been collected but not later than January 1, 1991. The tax will be phased out over a 33 month period. It is unlikely that the full $227.3 billion will be collected before the January 1, 1991 date.

In 1984 the tax was found unconstitutional by a federal district court in Wyoming. For an excise tax to be constitutional, it must be applied uniformly among all states. Certain Alaskan production, however, is exempt from the windfall profit tax. The Federal government successfully appealed the decision and the U.S. Supreme Court ruled that the WPT is constitutional.

The amount of the tax is based on a definition of windfall profit and a tax rate applied to that windfall profit. The amount of the windfall profit and the tax rate are related to the classification of the oil under the Emergency Petroleum Allocation Act of 1973 as amended. Thus, even though for oil pricing, oil is oil, when calculating the windfall profit tax, we can not forget the oil price classifications that were in effect under price controls. Generally, the category of oil established under price controls by the Department of Energy Regulations in effect on June 1, 1979, apply for WPT calculations.

10.1 CALCULATION OF TAX

The windfall profit tax is calculated by multiplying the ''windfall profit'' (IRC, Sec. 4988a) per barrel by the appropriate tax rate. The windfall profit is the removal price less an adjusted base price and an amount called the severance tax adjustment. The calculation can be expressed as:

$$WP = (RP - ABP - STA)$$
$$STA = (RP - ABP) * STR$$
$$WPT = WP * TR$$

where:

WP = windfall profit, $ per barrel
RP = removal (sales) price, $ per barrel
ABP = adjusted base price, $ per barrel
STA = severance tax adjustment, $ per barrel
STR = severance tax rate, decimal
TR = windfall profit tax rate, decimal
WPT = windfall profit tax, $ per barrel

The windfall profit per barrel of oil is limited to 90% of the ''net income'' attributable to the barrel. This is referred to as the net income limitation (NIL). An explanation of these terms as they apply to the windfall profit tax is presented in the following paragraphs.

10.1.1 *Removal Price*

The removal price (U.S. Treas. Reg. 150.4988-1(b)) is the amount for which the oil is sold including any adjustments that are made after the initial sale (BS&W and gravity corrections are two examples). A problem occurs when the oil is removed from the premises before it is sold. In this case the constructive sales price (IRC Section 613) used to determine the gross income for purposes of calculating depletion should be used. Transactions between related parties must also be made at a sales or removal price that is not less than the constructive sales price established by Section 613. In essence, the removal price should reflect the fair market value of the oil removed.

Oil that is used on the premises and is never removed from the premises is not subject to the windfall profit tax. For example, oil that is used to power a closed hydraulic pumping system is not considered to be sold for purposes of the windfall profit tax. Oil transferred from one property to another for purposes of powering production equipment is not taxed if each property has the same ownership.

10.1.2 *Base Price and Adjusted Base Price*

Base prices for a barrel of oil are established for three categories of production: Tier 1, Tier 2, and Tier 3 production. Production included in each category of oil is summarized below (U.S. Treas. Reg. 51.4991):

Tier 1 Oil: All domestic production not included in Tier 2 or Tier 3.

Production from Sadlerochit reservoir in Alaska.

Production from leases which produced during 1978.

Tier 2 Oil: Stripper well production. Any property where the average rate per producing well for any twelve consecutive month period after December 31, 1972 was ten barrels per day or less. Once classified as a stripper property, the property remains stripper property. Does not include oil that can be included in Tier 3 or, after December 31, 1982, oil that qualifies as exempt stripper oil. Exempt stripper oil is not "taxable crude oil" and accordingly is not included in any tier.

National Petroleum Reserve Production.

Condensate produced from non-associated gas production can not be classified as Tier 2 but is taxable crude oil for WPT purposes and must be included in Tier 1 or Tier 3.

Tier 3 Oil: Newly discovered oil. Oil that is produced from either an Outer Continental Shelf lease entered into after 1978, or an onshore property that did not have production during 1978.

Heavy oil. Oil with an API gravity of 16° or less.

Incremental Tertiary Oil. Incremental Tertiary Oil is the amount of oil produced above a base level from a qualified tertiary project.

A base price is established for each tier of oil on a property by property basis. The rules for calculating the base price for each tier are summarized below (U.S. Treasury Regulations, 51.4989). Once a base price is established for a category of oil produced from a property, the base price does not change even though the gravity or quality of the oil may change.

Tier 1 Oil: The base price for Tier 1 oil is the May 1979, DOE upper tier ceiling price, which averaged $13.02, less $0.21 ($12.81/bbl. average)

Tier 2 Oil: The Tier 2 base price is the uncontrolled base period price (December 1979) multiplied by the ratio 15.20/35.80. The national average price of uncontrolled crude oil in December, 1979 was deemed to be $35.80. Accordingly, the average base price for Tier 2 oil will be $15.20/bbl. The actual base price is determined on a property by property basis. The uncontrolled base period price is determined by comparing the sale of oil of similar gravity and quality from the nearest reservoir in cases where actual uncontrolled sales did not occur during the base period.

Tier 3 Oil: The base price for Tier 3 oil is the uncontrolled base period price multiplied by the ratio 16.55/35.80. Again, similar to Tier 2, the average Tier 3 base price would be $16.55. The actual base price is determined on a property by property basis in a manner similar to Tier 2 oil.

Once a base price is established, this base price is then adjusted by an inflation adjustment factor to determine the adjusted base price which is used to calculate the windfall profit.

The base price for Tier 1 and Tier 2 oil is adjusted for inflation based on the Gross National Product (GNP) implicit price deflator which measures inflation on domestically produced goods and services. There is a two quarter time lag in the inflation adjustment. For example, production that occurs in the fourth quarter of a year will have an adjusted base price increased for the inflation between the first and second quarter of the year. The inflation factor for Tier 3 oil is identical to the Tier 1 and Tier 2 factor except that the Tier 3 inflation factor is increased by an additional 2 percent nominal annual rate compounded quarterly. The following formula is used to determine the adjusted base price for the calendar quarter in which the oil is removed.

$$ABP = BP + BP * IAF$$

where:

ABP = Adjusted base price
BP = Base price
IAF = Inflation Adjustment Factor

For Tier 1 and Tier 2 Oil:

$$IAF = \frac{\text{GNP Implicit Price Deflator for the Quarter 2 quarters prior to the one in which oil was removed} - \text{GNP Implicit Price Deflator for the Second Quarter of 1979}}{\text{GNP Implicit Price Deflator for Second Quarter of 1979}}$$

For Tier 3 Oil:

$$IAF = \left[(1 + \text{Tier 1 IAF}) * (1.005)^{n}\right] - 1$$

where:

n = the number of quarters starting after September 30, 1979 but ending before the quarter in which the oil was removed

Inflation adjustment factors and the GNP Implicit Price Deflators are presented at the end of the chapter as Table 10-2.

10.1.3 *Severence Tax Adjustment*

The severance tax adjustment [IRC, Sec. 4996(c)] prevents the producer from paying windfall profit tax on the amount he pays to a state government as severance tax on the difference between the removal price and the adjusted base price. To qualify, the state severance tax must be based on the gross value of the crude oil. Severance taxes in excess of 15% also do not qualify for the severance tax adjustment. State severance taxes that are based on a fixed fee per barrel rather than the gross value of the produced oil do not qualify for the severance tax adjustment. Any changes in state severance taxes made after March 31, 1979 must apply to the entire price of a barrel of oil. The equation for calculating the severence tax adjustment is:

$$STA = (RP - ABP) * STR$$

where

STA	=	Severance tax adjustment
RP	=	Removal Price
ABP	=	Adjusted Base Price
STR	=	Severance tax rate

Table 9-1 in Chapter 9 presents a summary of severance tax rates. Some rates have been specifically accepted by the IRS as being eligible for the WPT severance tax adjustment, some have been specifically rejected for the severance tax adjustment, and some of the listed taxes have not been reviewed by the IRS. Accordingly, this table should only be used to estimate the severance taxes and WPT severance tax adjustment. Specific tax questions should be addressed to the IRS and the specific state involved.

10.1.4 *Net Income Limitation*

The calculated windfall profit should be compared to the net income income limitation (NIL) before the windfall profit tax is calculated (U.S. Treas. Reg. 51.4988-2). The NIL may reduce or eliminate the WPT on a marginal property. The net income limitation states that the windfall profit per barrel (for purposes of the WPT) is limited to 90% of the net income attributable to that barrel of oil. Thus the WPT is calculated using the lower of (1) the windfall profit, or (2) the NIL. Initial regulations required a strict matching of revenues and expenses for oil produced during a taxable year. To accomplish this matching of revenues and expenses attributable to taxable crude oil, income and expenses (cash and accrual) that occur during the current taxable year and two months (two month rule) after the close of the taxable year are to be included in the net income limitation for the taxable year if the income and expense are attributable to oil produced during the taxable year. A revised regulation was issued in October, 1981. This revised regulation eliminated the so called two month rule. Now taxable income for the net income limitation is based on the taxpayer's tax year and income tax method of accounting (i.e. accrual, cash). The net income limita-

---

Example 10-1
Calculation of the adjusted base price

1. Tier 1 oil having a base price of $12.81/bbl is produced in August, 1981.

   a) Calculate the inflation adjustment factor.

   $$IAF = \frac{1\ QTR\ 1981 - 161.17}{161.17}$$

   $$IAF = \frac{188.25 - 161.17}{161.17}$$

   $$IAF = 0.1680$$

   b) Calculate the adjusted base price

   $$\begin{aligned} ABP &= BP + BP * IAF \\ &= 12.81 + (12.81)*(0.1680) \\ &= \$14.96/bbl. \end{aligned}$$

2. Tier 3 oil having a base price of $16.50/bbl. is produced in August, 1981.

   a) Calculate the inflation adjustment factor.

   $$IAF = (1 + \text{Tier 1 IAF}) * (1.005)^n - 1$$

   $$\begin{aligned} IAF &= (1 + 0.1680) * (1.005)^7 - 1 \\ &= 1.2095 - 1 \\ &= 0.2095 \end{aligned}$$

   b) Calculate the adjusted base price.

   $$\begin{aligned} ABP &= BP + BP * IAF \\ &= 16.55 + 16.55 * 0.2095 \\ &= \$20.02/bbl. \end{aligned}$$

---

tion can be calculated using the following algorithm:

1. For the taxable year calculate gross income attributable to taxable crude oil from the property

2. Calculate allowable deductions for the property

   a. Operating expenses, losses, administrative and financial overhead, and depreciation for the property. Only costs attributable to taxable crude oil may be taken. Allocate costs not directly attributable to taxable crude oil, exempt oil, or gas production by the ratio of gross income from taxable crude oil to total gross income from the property.

   b. State and local taxes attributable to taxable crude oil for the property. The windfall profit tax may *not* be deducted.

   c. Imputed cost depletion (in lieu of expensing intangible drilling costs.

3. Calculate "net income" by subtracting the allowable deductions from the gross income (from taxable crude oil). The net income per barrel is the net income divided by the number of taxable barrels of crude oil sold during the tax year. The net income per barrel multiplied by 0.90 is then the net income limitation.

The imputed cost depletion calculation deserves further comment. Net income for purposes of calculating the net income limitation does not permit a deduction for intangible drilling costs or capitalized qualified tertiary injection expenses. However, these costs may be deducted through a pseudo or imputed cost depletion calculation. The depletable basis of a property as of the date of acquisition is determined. Subsequent capital expenditures are added to the basis. Cost depletion is calculated for each year (based on the estimate of reserves for that year) until the current tax year is reached. The allowable cost depletion for the current tax year is then allocated to taxable crude oil, exempt oil, and gas for the property. The amount allocated to taxable crude oil is an allowable deduction in calculating the net income limitation.

Example 10-2
Calculating the severance tax adjustment

1. The severance tax rate is 8% of the removal price of a barrel of oil. What is the severance tax adjustment per barrel if the removal price (RP) and adjusted base price (ABP) are \$35.00 and \$18.00 per barrel respectively.

$$\begin{aligned} STA &= (RP - ABP) * STR \\ &= (35.00 - 18.00) * 0.08 \\ &= \$1.36/\text{bbl.} \end{aligned}$$

In this case the windfall profit (WP) would be

$$\begin{aligned} WP &= RP - ABP - STA \\ &= 35.00 - 18.00 - 1.36 \\ &= \$15.64/\text{bbl.} \end{aligned}$$

2. The severance tax rate is 16% of the removal price. If the removal price and adjusted base price are the same as Part 1, what is the severance tax adjustment and the windfall profit per barrel.

$$\begin{aligned} STA &= STR * (RP - ABP) \\ &= 0.15\,(35.00 - 18.00) \quad (15\%\ \text{max}) \\ &= \$2.55/\text{bbl.} \end{aligned}$$

$$\begin{aligned} WP &= RP - ABP - STA \\ &= 35.00 - 18.00 - 2.55 \\ &= \$14.45/\text{bbl.} \end{aligned}$$

Example 10-3
Imputed Cost Depletion Calculation For WPT NIL

Two wells were drilled during 1981 on a property. One well was abandoned as a dry hole while the second well was completed as a producer. Revenue and cost information for 1981 and 1982 have been summarized below along with production and reserve information. Calculate the imputed cost depletion and the net income limitation for 1981 and 1982.

	1981 ($)	1982 ($)
Gross income from taxable crude oil @ $30/bbl.	150,000	900,000
Lease bonus	5,000	—
Dry hole costs	300,000	—
Intangible drilling costs	300,000	—
Tangible Costs	150,000	—
ACRS Depreciation of Tangible Costs	22,500	33,000
Operating Expenses	10,000	24,000
Severance Taxes	18,000	90,000
Allocated Overhead	10,000	15,000
Sales in barrels	5,000	30,000
Estimated End of Year Reserves (bbl)	50,000	60,000

1. A. Imputed Cost Depletion 1981 :

a) Calculate Depletable basis (refer to Chapter 11)

Lease bonus	$ 5,000
IDC's	300,000
	$ 305,000

b) Imputed Cost Depletion For 1981

$$CD = \text{Depletable Basis} * \frac{S}{S + U}$$

where:

S = Sales during the year
U = Estimated reserves at the end of the year.

$$\text{Imputed cost depletion} \quad \$305{,}000 \times \frac{5{,}000}{55{,}000} = \$3{,}182$$

Example 10-3
(Continued)

B. NIL For 1981 :

Gross Income from taxable crude oil @ $30/bbl.:		$ 150,000
Less:	Lease operating expenses	15,000
	Allocated Overhead	10,000
	Severance Taxes	18,000
	Dry hole costs	300,000
	Depreciation	22,500
	Imputed Cost Depletion	3,182
TAXABLE INCOME		$ <218,682>

Since the windfall profit tax is based on the lessor of (1) the windfall profit, or (2) the NIL, the windfall profit tax liability would be zero for 1981.

2. A. Imputed Cost Depletion for 1982

a) Depletable basis for 1982 = Depletable basis for 1981 less imputed cost depletion for 1981
= 305,000 − 3,182
= 301,818

b) $CD = 301{,}381 \times \frac{30{,}000}{90{,}000} = 100{,}606$

B. NIL 1982 For 1982:

Gross Income From Taxable Crude Oil @ $30/bbl.		$ 900,000
Less:	Lease Operating Expenses	24,000
	Allocated Overhead	15,000
	Severance Taxes	90,000
	Dry hole costs	—
	Depreciation	33,000
	Imputed Cost Depletion	100,606
TAXABLE INCOME		$ 637,394

$$NIL = \frac{0.90 * \$637{,}394}{30{,}000} = \$19.12/\text{bbl}$$

If the ''windfall profit'' per barrel before the NIL exceeds $19.12 for 1982, the tax will be based on $19.12/bbl.

10.1.5 *Tax Rates*

The tax rate (IRC, Sec. 4987) that is applied to the calculated windfall profit is based on the crude oil classification (Tier 1, Tier 2, or Tier 3), and the status of the economic interest owner (major, independent, or royalty). Table 10-1 presents the appropriate tax rate for each tier of oil and type of ownership. Subsequent paragraphs will present the qualifications for the reduced independent producer rates.

Since the status of the economic interest owner is needed to determine the windfall profit tax rate, three economic interest owners will be discussed: Independent producers, major oil producers, and royalty owners.

The *independent producer* (IRC, Sec. 4992) has a lower tax rate applied to its first 1000 barrels per day of Tier 1 and Tier 2 oil. The independent producer must have a working interest ownership, must have had less than $1,250,000 in retail sales of oil, gas, and derivatives of oil and gas in the prior calendar quarter, and must have had less than 50,000 BPD of refinery runs for every day during the prior calendar quarter. To qualify for the reduced independent producer rates, the working interest ownership must have been in effect on January 1, 1980. Overriding royalty interests on January 1, 1980 having a binding contract in existance by February 20, 1980 to permit conversion of the overriding interest to a working interest ownership may qualify for the reduced independent producer tax rate when the conversion to a working interest ownership is made.

Rules concerning related parties should be considered when determining if an entity is disqualified for the reduced independent producer rate because of retail sales or refinery runs.

The amount of production qualified for the reduced rate is calculated on a quarterly basis. If the quarter consists of 90 days, then the qualified production would be 90,000 barrels. If the daily average production exceeds 1000 barrels per day, the production is allocated between Tier 1 and Tier 2 oil in proportion to the producer's total amount of Tier 1 and Tier 2 production. The Tier 1 or Tier 2 production that exceeds the allocated amount is taxed at the higher rate. Within a Tier, barrels sold at the highest price receive the reduced tax rate first. This concept is illustrated by Example 10-5.

Royalty ownership is important since certain exemptions apply. Otherwise, royalty oil is taxed like a major oil producer. A royalty owner has an economic interest in production from a property but does *not* share in the costs of produc-

---

Example 10-4
Windfall Profit Tax Net Income Limitation

Determine the windfall profit for the following property.

Removal price:	$35.00/bbl
Adjusted base price:	$18.00/bbl
Severance tax adjustment:	$ 1.50/bbl
Net income:	$16.00/bbl

Windfall profit before net income limitation

WP = 35.00 — 18.00 — 1.50
= $15.50/bbl

Net income limitation

NIL = 0.90 x $16.00
= $14.40/bbl

Therefore the windfall profit is limited to $14.40/bbl for calculating the windfall profit tax.

---

tion. A royalty owner also does not have any right to explore for minerals on the property. The royalty owner is only entitled to a share of the minerals produced or the proceeds from the sale of the produced mineral.

Royalty owners received a $1000 credit against windfall profit tax liability in 1980. The credit was increased to $2500 in 1981. The credit claimed can not exceed the actual liability incurred. For 1982 and subsequent taxable years the credit is replaced with an exemption. From 1982 through 1984, the royalty owner may exempt 2 barrels per day of qualified royalty crude oil production. Starting in 1985, three barrels per day of qualified royalty crude oil production will be exempt. Qualified royalty production, defined in the IRC Section 6429, has certain limitations.

A. Qualified royalty owners production does not include taxable crude oil attributable to any overriding royalty interest, production payment, net profits interest, or similar interest that is created after June 9, 1981 out of an operating mineral interest in property which is a proven oil or gas property on the date the interest is created, or was not created pursuant to a binding contract entered into prior to June 10, 1981.

B. Qualified royalty owners production does not include production from a property transferred after June 9, 1981 unless the the production qualified for the royalty credit before the transfer, and the transferees share the royalty credit.

C. Subleases are considered transfers of property, and

Table 10-1
WINDFALL PROFIT TAX RATES
(Incorporates 1984 Tax Reform Act)

		INDEPENDENT (1000 B/D Limit)	MAJOR & ROYALTY
TIER 1			
All Years		50%	70%
TIER 2			
All Years		30	60
TIER 3			
All Years Except Newly Discovered		30	30
Newly Discovered:			
Before	1982	30.0	30.0
	1982	27.5	27.5
	1983	25.0	25.0
	1984	22.5	22.5
	1985	22.5	22.5
	1986	22.5	22.5
	1987	22.5	22.5
	1988	20.0	20.0
	1989	15.0	15.0
After	1989	15.0	15.0

Example 10-5
Qualified Independent Producer Production

Assume all of the following production applies to the working interest ownership of a qualified independent producer.

	Sales 1st Qtr 1981 (bbl)	Price ($/bbl)
Tier 1	80,000	10
Tier 1	25,000	16
Tier 1	10,000	38
Tier 2	40,000	38

Independent producer limit:

$$1000\ \frac{\text{Bbl}}{\text{Day}} \times 90\ \frac{\text{day}}{\text{Qtr}} = 90{,}000 \text{ Bbl per quarter}$$

Allocated Independent Producer Amount:

For Tier 1:

$$\frac{\text{Tier 1 production}}{\text{Total Tier 1 and Tier 2}} \times \text{Independent Producer Limit}$$

$$= \frac{115{,}000}{155{,}000} \times 90{,}000 = 66{,}774 \text{ bbls}$$

For Tier 2:

$$\frac{\text{Tier 2 Production}}{\text{Total Tier 1 and 2 Production}} \times \text{Independent Producer Limit}$$

$$= \frac{40{,}000}{155{,}000} \times 90{,}000 = 23{,}226 \text{ bbls}$$

Since the Tier 1 production sales are at various prices, the production with the highest sales price within Tier 1 is used first.

	66,774	bbl. Tier 1 qualify for reduced rate
Less	10,000	bbl. Tier 1 sold at $38/bbl
	56,774	
Less	25,000	bbl. Tier 1 sold at $16/bbl
	31,774	bbl. Tier 1 sold at $10/bbl

Therefore, 10,000 bbl sold at $38/bbl, 25,000 bbl sold at $16/bbl, and 31,774 bbl sold at $10/bbl are eligible for the reduced independent producers rate. The remaining (80,000 − 31,774) 48,226 bbls of Tier 1 oil are taxed at the higher rate.

Tier 2 production is all at the same price. Therefore, 23,226 bbl of Tier 2 oil priced at $38/bbl qualify for the reduced independent producers rate. The remaining (40,000 - 23,225) 16775 barrels are taxed at the higher rate.

D. Property held by an estate is considered owned by both the estate and the beneficiaries of the estate.

A *major oil producer* is a working interest owner that does not qualify for the reduced independent producer rate because of retail sales or refinery runs. A major is taxed at the higher tax rate.

10.2 EXEMPT OIL AND TRANSFER RULES

Two other features of the WPTA will be discussed: (1) exempt oil, and (2) transfer rules.

10.2.1 *Exempt Oil*

All domestic crude oil is subject to the windfall profit tax unless specifically exempted. The paragraphs which follow summarize exempt oil (IRC, Sec. 4994 and Federal Tax Guide, p. 2457-3 and 2458):

1. Oil produced by state and local governments is exempt. The proceeds, however, must be used for public purposes.

2. Oil produced by charitable, medical or educational institutions, or a residental child care agency is exempt. Interest must have been owned on January 21, 1980 and held through the taxable year.

3. Alaskan oil produced from wells north of the Arctic Circle is exempt. Additionally, oil produced north of the divides of the Alaskan and Aleutian Mountain Ranges and south of the Arctic Circle is exempt if the producing well is more than 75 miles from the nearest point on the Trans-Alaska Pipeline. Oil produced from the Sadlerochit Reservor is taxed as Tier 1 oil.

4. Oil produced by federally recognized Indian tribes is exempt. Interests owned by or on behalf of Indian tribes, tribal organizations, or individual Indians on January 21, 1980 is exempt.

5. Starting in 1982, royalty owners may exempt two barrels per day through 1984. Starting in 1985, three barrels per day of crude oil will be exempt.

6. Certain stripper oil produced after 1982 is exempt. The following four conditions must be met.

   a. oil must be removed from a stripper well property after 1982.

   b. oil must be produced by an independent producer.

   c. only applies to independent producers working interest.

   d. the property may not be transferred by a non-independent producer on or after July 23, 1981.

10.2.2 *Transfer Rules*

A transfer is a sale, exchange, lease, sublease, gift or bequest of rights with respect to the property (U.S. Treasury Regulations, 150.4996-3). For property transfers after 1978 to be classified as stripper, newly discovered, or heavy oil after the transfer, they must have been classified as such before the transfer.

In general, on or after January 1, 1980 a *transferred property* is only eligible for the reduced independent producer's rate if the transferor was an independent who did not produce in excess of 1000 barrels of oil per day in any quarter after September, 1979 including the quarter before the quarter in which the transfer is made. Therefore, only production transferred between two independents (where the transferor's production never exceeds the 1000 barrel per day limit in any quarter) qualifies for the reduced independent producer's rate. The Internal Revenue Code and regulations should be checked in detail to ensure that a property transfer qualifies for the reduced independent producer's rate.

Table 10-2

INFLATION ADJUSTMENT FACTORS
AND
GNP IMPLICIT PRICE DEFLATORS

Calendar Quarter	Inflation Adjustment Factors: Tier 1 and Tier 2	Inflation Adjustment Factors: Tier 3	GNP Implicit Price Deflator
2nd Qtr 1979	NA	NA	161.85*
3rd Qtr 1979	NA	NA	166.99
4th Qtr 1979	NA	NA	170.71
1st Qtr 1980	0.0195	0.0246	174.42
2nd Qtr 1980	0.0422	0.0527	178.93
3rd Qtr 1980	0.0649	0.0810	183.23
4th Qtr 1980	0.0924	0.1144	183.80
1st Qtr 1981	0.1187	0.1469	188.25
2nd Qtr 1981	0.1404	0.1750	191.17
3rd Qtr 1981	0.1680	0.2095	195.44
4th Qtr 1981	0.1861	0.2344	200.01
1st Qtr 1982	0.2126	0.2683	201.84
2nd Qtr 1982	0.2410	0.3045	206.14
3rd Qtr 1982	0.2523	0.3230	208.38
4th Qtr 1982	0.2736	0.3522	210.42
1st Qtr 1983	0.2875	0.3737	213.36
2nd Qtr 1983	0.3001	0.3941	214.66
3rd Qtr 1983	0.3183	0.4207	216.29
4th Qtr 1983	0.3263	0.4365	218.67
1st Qtr 1984	0.3364	0.4546	220.52
2nd Qtr 1984	0.3511	0.4780	222.33
3rd Qtr 1984	0.3625	0.4979	224.44
4th Qtr 1984	0.3737	0.5178	226.10
1st Qtr 1985	0.3867	0.5398	
2nd Qtr 1985	0.3970	0.5590	
3rd Qtr 1985			
4th Qtr 1985			

* Since the WPT act was passed, the GNP implicit price deflator for the 2nd Quarter of 1979 has been revised twice. To calculate the inflation adjustment factor for quarters preceding the 2nd Quarter of 1981, use 163.79. Beginning with the 2nd Quarter of 1981 but preceding the 4th Quarter of 1982, use 161.17.

References

Crude Oil Pricing/Windfall Profit Tax Information Service, Federal Programs Advisory Service, Thompson Publishing Group, Washington, D.C.

Federal Tax Guide, 1983, Commerce Clearing House, 4025 W. Peterson Ave., Chicago, Illinois.

Internal Revenue Code, 1954.

U.S. Treasury Regulations.

PROBLEMS

10-1 Calculate the windfall profit tax per barrel for production during the 2nd quarter of 1982 for each tier of oil and producer status. The removal price is $30 per barrel, the qualifying severance tax rate is 10% of the gross value. The base prices are summarized in the following table.

	Base Price
Tier 1	12.81
Tier 2	15.20
Tier 3	16.55

10-2 A qualified independent producer sells the following amount of oil attributable to its working interest during the 3rd quarter of 1983:

Tier 1	80,000 bbl
Tier 2	20,000 bbl

a. How much of the production (barrels) qualifies for the reduced independent producers rate?

b. Allocate the production qualifying for the reduced independent rate between Tier 1 and 2. All production is sold at the same price.

10-3 An old field that has produced since 1936 is shut-in during 1977. No production occurs during 1978. However, as a result of price decontrol, production is commenced again in 1979.

a. How would you classify this oil; Tier 1, Tier 2, or Tier 3?

b. For sales during the 3rd quarter of 1983, what is the windfall profit tax rate?

c. In part b, if the base price of oil is $17.00 per barrel, the removal price is $28.00 per barrel, and the severance tax rate is 8% of gross revenue, calculate the windfall profit tax per barrel. Ignore the net income limitation.

d. If the net income per barrel is $3.80, calculate the net income limitation.

e. Considering the net income limitation, what is the windfall profit tax per barrel (use your answers from part b, c, and d).

Table of Contents
Chapter 11
Federal Income Taxation of Oil and Gas Transactions

11.1 Simplified Tax Model 11-1
11.1.1 Gross Income 11-1
11.1.2 Allowable Deductions 11-2
11.1.3 Taxable Income 11-2
11.1.4 Carryovers and Carrybacks 11-2
11.1.5 Tax Credits 11-2
11.1.6 Minimum Tax 11-2
11.2 Economic Interest Concept 11-2
11.2.1 Working Interest 11-3
11.2.2 Royalty Interest 11-3
11.2.3 Overriding Royalty Interest 11-3
11.2.4 Net Profits Interest 11-3
11.2.5 Carried Interest 11-3
11.2.6 Production Payments 11-4
11.3 Property Concept 11-4
11.3.1 Separate Tract 11-4
11.3.2 Separate Deposit 11-4
11.3.3 Separate Interest 11-5
11.4 Definitions of Common Oil and Gas Transactions 11-5
11.4.1 Leasehold Costs 11-5
11.4.1.1 Lease Bonus 11-5
11.4.1.2 Geological and Geophysical Costs 11-5
11.4.2 Lease and Well Equipment 11-6
11.4.3 Intangible Drilling and Development Costs 11-6
11.4.4 Lease Operating Expense 11-6
11.5 Federal Income Taxation Model 11-6
11.5.1 Gross Income 11-6
11.5.2 Allowable Deductions 11-6
11.5.2.1 Delay Rentals 11-7
11.5.2.2 Operating Costs 11-7
11.5.2.3 Taxes 11-7
11.5.2.4 Interest Expenses 11-7
11.5.2.5 Allocated Overhead 11-7
11.5.2.6 Intangible Drilling Costs 11-7
11.5.2.7 Depreciation 11-10
11.5.2.7.1 Classes of Property 11-10
11.5.2.7.2 Calculating ACRS Depreciation 11-10
11.5.2.8 Allowable Depletion 11-12
11.5.2.8.1 Cost Depletion 11-12
11.5.2.8.2 Percentage Depletion 11-13
11.5.2.8.2.1 Certain Natural Gas 11-13
11.5.2.8.2.1 IPRO Production 11-13
11.5.2.8.3 Depletion Allowance Calculation 11-15
11.5.3 Investment Tax Credit 11-16
11.5.4 Minimum Tax 11-20
11.5.4.1 Tax Preference Items 11-20
11.5.4.1.1 Depletion 11-20
11.5.4.1.2 Intangible Drilling Costs 11-21
11.5.4.2 Noncorporate Taxpayers 11-22
11.5.4.3 Corporate Taxpayers 11-23

Chapter 11

Federal Income Taxation of Oil and Gas Transactions

The tax implications of oil and gas transactions are by no means simple. Tax laws are passed by Congress which has the ultimate authority to levy and collect taxes. The passage of tax laws is a process of compromise which often alters proposed legislation to such a degree that it is sometimes difficult to determine the "why" of certain features of a tax. Our objective, however, is not to explain why, but rather to present the tax treatment of common oil and gas transactions. We do not propose to present ourselves as experts in the area of oil and gas taxation. Specific tax questions should be directed to a tax professional. This chapter establishes the framework for the development of a Federal income tax model that can be used to estimate net cash flow after Federal income taxes. The remainder of the chapter will show the relationship between our Federal income tax model and the cash flow model for oil property evaluation developed in Chapter 1, and the application of tax concepts to oil and gas transactions.

11.1 SIMPLIFIED TAX MODEL

First let's consider the following simplified tax model.

Gross Income
− Allowable Deductions
= Taxable Income
× Effective Tax Rate
= Tax Liability Before Credits and Minimum Tax
− Tax Credits
+ Minimum Tax on Items of Tax Preference
= Cash Taxes

A negative "cash tax" is treated as an item of cash *inflow* in the cash flow model. This assumes that either sufficient tax liability exists from other activities of the entity or that the net operating loss or tax credit can be carried back. Carryovers and carrybacks will be discussed later in the chapter. A positive cash tax is treated as an item of cash outflow in the cash flow model.

To develop an oil and gas model for the taxation of oil and gas transactions, we must first understand some basic terminology used in taxation. The following paragraphs will present the meaning of gross income, allowable deductions, taxable income, tax carryovers and carrybacks, and tax credit in the context of income taxation. It is important to remember that although terminology such as depreciation and depletion (allowable deductions) are used in both accounting and taxation, the definitions are not the same.

11.1.1 *Gross Income*

Gross income is "gain derived from capital or labor". For income tax calculations, income is generally defined as all income except items specifically excluded by law. The form of the income is not important. Income may be in the form of services, meals, stock, property, or cash. The Internal Revenue Code (IRC) Section 61(a) lists the following examples of income:

(1) Compensation for services, including fees, and commissions;
(2) Gross income derived from business;
(3) Gains derived from dealings in business;
(4) Interest;
(5) Rents;

(6) Royalty; and
(7) Dividends.

The above list is by no means all inclusive, and is only used to illustrate some of the items that are included in gross income.

11.1.2 *Allowable Deductions*

Allowable deductions are items that the tax code specifically allows to be subtracted from gross income to arrive at taxable income. Thus, deductions reduce the amount of income that is subject to taxes. Examples of allowable deductions from gross income are operating costs, state and local taxes, windfall profit tax, depreciation of tangible equipment, and depletion of depletable items such as leasehold costs. Oil and gas transactions involve many special rules in the area of capital expenditures. For example, the taxpayer may elect to expense or capitalize some items such as intangible drilling costs.

11.1.3 *Taxable Income*

Taxable income is used to compute a taxpayers tentative tax liability. For our purposes we will assume an effective tax rate that will be applied to the entire taxable income figure. The tax liability is only tentative since tax credits and the minimum tax on tax preference items have not been considered.

11.1.4 *Carryovers and Carrybacks*

Certain items (such as net operating losses and investment tax credits) may, under certain conditions, be carried forward (carryovers) to subsequent years or carried back to prior years. Generally, items that qualify can be carried back 3 years and forward 15 years. Since the tax liability associated with a project is an incremental analysis, the tax for the project is the tax liability of the company with the project less the tax liability of the company without the project. In the case where your project generates a tax savings for the company but there is a negative cash tax for the company (with your project), the possibility of a carryback to prior years of either the net operating loss or investment tax credit that caused the negative cash tax needs to be examined. If the items can be carried back according to the tax code and reduce the amount of cash taxes paid in the carryback year, the project can be given credit for the tax savings as a positive cash flow in the current tax year. The project cash flow can only reflect the portion of the tax savings attributable to the project that can be used in the current taxable year. The tax savings (negative cash tax) that cannot be carried back or used in the current taxable year should be analyzed for the possibility of a carryover. If the tax savings is the result of an item that can be carried forward to subsequent years, then the project's cash flow should reflect a positive cash flow for taxes in the taxable year in which the tax savings can be used. For project analysis this obviously involves an estimate of the future activities of the firm.

11.1.5 *Tax Credits*

Tax credits are applied directly to the tentative tax due (taxable income times the effective tax rate). A credit reduces the tax liability on a dollar for dollar basis.

11.1.6 *Minimum Tax*

Minimum tax calculations are performed to determine if the taxpayer owes additional taxes because of the use of certain allowable deductions and exclusions called items of tax preference. The concept behind the minimum tax calculation is that taxpayers should not be able to avoid all taxes through the use of items of tax preference. In other words, the tax law is saying that you can only have so much of a good thing, then enough is enough. The minimum tax calculation is different for individuals and corporations and will be presented later. Two examples of items of tax preference in the oil business are certain intangible drilling costs and percentage depletion.

11.2 ECONOMIC INTEREST CONCEPT

In addition to the basic terms presented above, there are two major concepts that are pervasive throughout oil and gas tax calculations: the *economic interest* and *property* concepts. These two concepts will be discussed before we consider a more detailed model of federal income taxation.

The concept of an economic interest is important in taxation because it is used to characterize whether or not the taxpayer's income is subject to depletion. Since the concept of depletion and the depletion allowance have not been discussed, only what constitutes an economic interest in oil and gas production will be presented here. Depletion calculations will be presented later in the chapter.

The U. S. Treasury Regulation Sec. 1.611-1(6)(1) gives the following definition of an economic interest.

> An economic interest is possessed in every case in which the taxpayer has acquired by investment any interest in mineral in place . . . and secures, by any form of legal relationship, income derived from the extraction of the mineral . . ., to which he must look for a return of his capital.

Generally an economic interest exists where the taxpayer has a working interest, royalty interest, overriding royalty interest, and net profits interest. The following paragraphs describe the various types of economic interests.

11.2.1 *Working Interest*

The working interest owner has an economic interest in the minerals in place and must look to that interest for a return of his investment. The working interest owner must pay all development and operating costs. A working interest owner must take into account the cost of operating the property when calculating the taxable income limitation for percentage depletion. A working interest is said to be a depletable interest since the working interest owner can take depletion as an allowable deduction in calculating taxable income.

11.2.2 *Royalty Interest*

A royalty interest owner has the right to a share of production, or the value of the production, without any responsibility for the cost to develop or operate the property. A royalty interest is created by the original owner of the minerals.

11.2.3 *Overriding Royalty Interest*

An overriding royalty interest, similar to a royalty interest, has a share of the production without any responsibility for development or operating costs.

11.2.4 *Net Profits Interest*

A net profits interest is an economic interest in the production that entitles the holder to a specific fraction of the profit as defined by the contract. Profit usually is net of development, operating costs, and production taxes. The net profits interest is an economic interest that may take the depletion allowance in calculating federal income taxes. Unlike a working interest owner, the net profits interest is not liable for costs in excess of those that can be recovered through production.

11.2.5 *Carried Interest*

A carried interest transaction may take many forms. A typical situation is where two parties, one the carrying party and the other the carried party, make an arrangement whereby the carrying party advances all or part of the development costs for a property in return for a share of the proceeds of any production from the property. Usually on payout, the carrying party earns a portion of the carried party's working interest in the well. Three major tax cases referred to as Abercrombie, Manahan, and Herndon, have guided the tax courts in establishing the tax consequence of carried interest sharing agreements. More information on these cases can be found in *Oil and Gas Law* (Williams and Meyers, 1981) and *Income Taxation of Natural Resources* (Burke and Bowhay, 1983). The Manahan Case is now used to define the tax consequences of most carried interests agreements. Here the carrying party may expense the same fraction of the total IDC's as the fraction of the working ownership that the carrying party can look to for a return of his investment during the payout period. U. S. Treasury Regulation Section 1.612-4(a) provides that intangible drilling costs and depreciable costs attributable to interests held by others (the carried party) must be capitalized and depleted by the carrying party rather than having the option to expense the intangible drilling costs (IDC's) and capitalize and recover through depreciation the tangible property. In other words, if the carrying party pays for 100% of the costs to drill and develop a well, then it can expense 100% of its IDC's as long as it can look to 100% of the carried parties interest for the full return of its investment (until payout). However, if the carrying party pays 100% of the costs, but can look only to a 70% working interests share of revenue for a return of his investment, the carrying party can only expense 70% of the IDC's and must capitalize and recover through depletion the remaining 30%. Since the structure of the deal may have significant tax consequences, it is important to seek professional advice in this area. After payout, the terms may be any amount agreed to in the contract.

11.2.6 *Production Payment*

A production payment is the right to minerals in place that gives the owner of the production payment the right to a specified fraction of the production until a certain point in time is reached or until a certain amount of production has been accrued. Prior to August 7, 1969 all production payments were considered economic interests. Subsequent to this date, production payments must meet *either* of the following conditions to be classified as an economic interest (IRC Section 636):

1. A production payment carved out of a mineral property where the proceeds are for the exploration and development of the property.
2. A production payment retained in a mineral property by the lessor in a leasing transaction.

11.3 PROPERTY CONCEPT

The property concept is an important tax consideration in almost every oil and gas transaction. Gross income, depletion, geologic and geophysical costs, intangible drilling costs, depletion allowance, gain or loss on sale of a property and the resulting tax consequences are all calculated on a property by property basis. For this reason the following paragraphs will establish some important variations that are possible in establishing the property unit for federal income tax purposes. Special rules apply for properties held before 1954, for properties acquired between 1954 and 1963, and for properties acquired after 1963. Smith, Anderson, and Jones (1982) discuss the property concept.

U. S. Treasury Regulation Section 1.614-1(a) defines a property as "... each separate interest owned by the taxpayer in each mineral deposit in each separate tract or parcel of land." This definition has three separate parts:

(1) each separate tract or parcel of land,
(2) each separate deposit, and
(3) each separate interest.

The analysis of the property concept is not an easy one. Our objective here is to give the reader an idea of the nature of the property concept so that he will at least know the importance of seeking professional guidance in this area. The concepts presented apply to properties acquired after December 31, 1963.

11.3.1 *Separate Tract*

The separate tract or parcel of land concept simply applies to the physical surface description of the land in the lease agreement or other instrument of conveyance. If the conveyance describes more than one tract, then consideration is given to the location of the tracts relative to one another. If the tracts are all one contiguous area then the tracts will be considered one property. If the tracts are connected only by a common corner, then each tract will be a separate property. Tracts that are geographically separated are separate properties.

Contiguous tracts obtained from different grantors are considered to be separate properties.

11.3.2 *Separate Deposit*

Separate mineral deposits underlying a single tract may also constitute a separate property if the taxpayer elects to do so. Separate, distinct, geologic formations qualify as a property under the property concept. The geologic horizons must be clearly identifiable and may not be merely many producing zones in the same formation. If four distinct deposits underlie a parcel of land, then (if there is a single interest in the parcel) there may be four separate properties if the taxpayer elects.

11.3.3 *Separate Interest*

A separate interest for the property concept has the same meaning given to the economic interests which were presented earlier. Working, royalty, overriding royalty, and net profits interests are included. Under the separate interest concept, a separate tract of land defined as a property earlier may be more than one separate property when the separate interest concept is applied. This can occur if the taxpayer has both a working interest and a royalty interest, for example. For application of the separate interest concept, the economic interests presented earlier must be classified as either an *operating* or *nonoperating* interest. The IRC Section 614(d) defines an operating interest as an interest that requires the cost of production to be taken into account when calculating the 50 percent limitation for percentage depletion. All other interests are nonoperating. A working interest is an operating interest; royalty, overriding royalty, and net profits interests are nonoperating interests.

If a taxpayer has two or more *operating* interests on the same tract or parcel of land, the interests must be combined into a single property unless the taxpayer elects not to do so. If the taxpayer has two or more *nonoperating* interests on the same tract, the taxpayer may treat the interests as a single property if the taxpayer can establish that tax avoidance is not the reason for the aggregation. Operating interests generally must be kept separate from nonoperating interests for the property concept.

11.4 DEFINITIONS OF COMMON OIL AND GAS TRANSACTIONS

The tax treatment of oil and gas transactions will be discussed using the Federal Income Tax Model, Figure 11-1, as the various transactions appear in the tax model. First, however, certain tax terminology used in the oil and gas industry will be established.

11.4.1 *Leasehold Costs*

Leasehold costs are expenditures associated with the acquisition of an economic interest in a natural resource that are deemed to benefit future tax periods. Accordingly leasehold costs are *capitalized* and the tax free return of that expenditure recovered through *depletion*. Two important leasehold costs are the lease bonus and geologic and geophysical costs.

11.4.1.1 *Lease Bonus*

In leasing transactions, the payment made by the lessee to the lessor for the right to explore for oil and gas and develop the property is called a lease bonus. Typically the lease bonus gives the lessee the right for a one year period to explore for oil and gas and to develop the property. If production has not been established by the end of the first year or drilling has not commenced, then the lease often provides for a payment called a *delay rental* that must be paid to extend the lease for another year. Typically, delay rentals are treated as expenses versus being capitalized. The process of paying delay rentals to extend the lease may continue until the primary term of the lease expires.

11.4.1.2 *Geological and Geophysical Costs*

Geological and geophysical (G & G) costs are expenditures for geologic studies and geophysical work such as seismic surveys, that are incurred to find and delineate drilling prospects. When G & G costs result in the acquisition and retention of a property, the expenditures should be *capitalized* and recovered through *depletion*. When G & G costs do not lead to the acquisition or retention of a property, then the G & G costs can be *expensed* in the taxable year.

In allocating G & G costs two terms are used; *project costs* and *area of interest costs*. Project costs are allocated to a project area. If an operator can conduct continuous operations in an area, the area is deemed to be a project area. Usually a reconnaissance survey is performed to determine areas of interest within the project area. Project costs are allocated to each area of interest based on acreage. Each area of interest is then surveyed in detail. Area of interest costs are the direct costs incurred plus the previously allocated project costs distributed on a net acreage basis. For retained acreage these costs establish the capitalized G & G expenditures for each property. In the areas of interest where no property is retained, the G & G costs are written off as a loss.

11.4.2 *Lease and Well Equipment*

Lease and well equipment includes items such as the wellhead, flow lines, separators, production tanks, and other equipment necessary to operate the property as well as the labor to install the equipment. *Tangible* property such as this is recovered through *depreciation*. Subsequent to 1980, the accelerated cost recovery system (ACRS) of capital recovery must be used. This system and classification of property are discussed with the Federal Income Tax Model.

11.4.3 *Intangible Drilling and Development Costs*

Special tax treatment for expenditures classified as intangible drilling and development costs or IDC's is available to the taxpayer. The U.S. Treasury Regulation 1.612-4 states that ''all expenditures made by an operator for wages, fuel, repairs, hauling, supplies, etc., incident to and necessary for the drilling of wells and the preparation of wells for the production of oil or gas'' are IDC's. A well is considered to be prepared for production when the wellhead (Christmas tree) is installed. *Tangible* items such as surface casing and production casing are classified as tangible expenditures and are recovered through *depreciation*. At the election of the taxpayer the IDC's may be either expensed or capitalized and recovered through depreciation and depletion. Specifics about the election to expense or capitalize and the tax consequences are presented with the discussion of the Federal Income Tax Model.

11.4.4 *Lease Operating Expense*

Lease operating expenses are expenditures incurred in the day to day activities of the production operations. Basically, costs incurred during the current tax period that were spent in an effort to generate revenues during the current tax period are treated as current period expenses. That is, the expenditure will benefit only the current period and thus is *not* capital in nature. Some example operating costs include direct labor, power costs, and maintenance costs.

11.5 FEDERAL INCOME TAXATION MODEL

The following discussion of oil and gas taxation will be based on the Federal Income Tax Model For Oil and Gas Transactions presented as Figure 11-1. All amounts in the model are net numbers (to your working and revenue interest) and are for your taxable period which is usually one year.

11.5.1 *Gross Income*

Gross income is calculated on a property by property basis. This is necessary since the percentage depletion calculations presented later are made for each individual property. U. S. Treasury Regulation Section 1.613-3 defines gross income from a property as ''the amount for which the taxpayer sells the oil or gas in the immediate vicinity of the well. If the oil or gas is not sold on the premises but is manufactured or converted into a refined product prior to sale, or is transported from the premises prior to sale, the gross income from the property shall be assumed to be equivalent to the representative market or field price of the oil or gas before conversion or transportation.'' Only the income from a taxpayer's economic interest is included in gross income. For example, royalties paid are not included in gross income.

Simply, gross income from a particular property would be *your* sales volume times the sales price per unit.

11.5.2 *Allowable Deductions*

Taxable income before a depletion allowance must be calculated for each property in order to compute percentage depletion. Certain items called allowable deductions are subtracted from gross income to compute taxable income. Allowable deductions include charges directly related to the property but also include administrative and overhead charges that are properly allocated to the property. The following paragraphs discuss the tax treatment of specific allowable deductions.

11.5.2.1 *Delay Rentals*

Delay rentals are usually annual payments made by the lessee to the lessor to keep or maintain a lease during the primary term of the lease until drilling commences or production is established.

Delay rentals may be either expensed or capitalized by the lessee at the election of the taxpayer. The election can be made annually on a property by property basis. Usually delay rentals are expensed. However, some tax situations may dictate that the delay rental should be capitalized and recovered through depletion. One example is where a net operating loss (NOL) is expected to occur and the NOL can not be carried forward or back. Thus, by capitalizing the delay rental you will reduce your taxable income in subsequent years. The gain derived from putting the expense in future years may not be realized if the taxpayer is eligible for percentage depletion. This should become clear after percentage depletion is discussed later in the chapter.

11.5.2.2 *Operating Costs*

Operating costs (lease operating expenses) are expenses incurred in operating the property and include the direct labor and supplies that are needed to produce the oil and gas. Examples of operating costs include maintenance of equipment, workover costs, and power costs for pumping units. Direct operating costs incurred to maintain reservoir pressure by injecting water into the reservoir are also operating costs.

11.5.2.3 *Taxes*

Severance and advalorem taxes as well as the windfall profit tax attributable to a property must be deducted in computing taxable income before depletion. Some income taxes must be deducted in computing taxable income before depletion if the tax is attributable to income derived from the property. The income tax in the State of Oklahoma is an example of such an income tax.

11.5.2.4 *Interest Expenses*

Interest paid on loans made to purchase a producing property are deducted from the gross income for that property.

11.5.2.5 *Allocated Overhead*

Some expenditures such as officer salaries, office expenses, legal expenses, and general taxes are not directly attributable to a property but are considered ordinary and necessary business expenditures. If the only business of the taxpayer is oil and gas production, then these overhead items must be allocated among each property. However, if the taxpayer conducts business in other areas, then the overhead charges must first be allocated among the business activities and then among the oil and gas properties. Allocation must be made in a reasonable and consistent manner. Often allocation in proportion to direct expenses is used to allocate expenses among the various business activities. Production is sometimes used to allocate overhead among various producing properties.

11.5.2.6 *Intangible Drilling Costs*

IDC's are computed on a property basis for the economic interest owners. Intangible drilling costs (IDC's) may, at the option of the taxpayer, be deducted as an expense in the current tax year or capitalized. Tangible items such as surface casing (even if cemented in the well without apparent salvage value) and production casing are capitalized and recovered through depreciation. Table 11-1 presents the classification, IDC or tangible, of typical expenditures incurred in preparing a well for production. Burke and Bowhay (1983, p. 1421-1422) have additional examples.

The statement electing to expense or capitalize IDC's should be made with the first return after a well is drilled even if the return reflects only dry hole costs for a nonproductive well. If a taxpayer elects to capitalize IDC's, he may still elect to expense dry hole costs in the tax year the well is abandoned and declared unproductive. Thus, the act of expensing a dry hole in the first tax return for a property would not imply the intention of the taxpayer to expense future IDC's. If

FIGURE 11-1

Federal Income Tax Model For Oil and Gas Transactions

Amounts are Net to Your Working and Revenue Interest

GROSS INCOME

Units Sold x $\frac{\$}{\text{Unit}}$

less delay rentals (may elect to capitalize)

less operating costs

less state and local taxes, windfall profit tax

less interest on loan

less allocated overhead

less expensed IDC's (Figure 11-2)

less dry hole costs (Figure 11-2)

less depreciation (Figure 11-3)

DEPRECIABLE ITEMS
lease and well equipment (Tangibles)

TAXABLE INCOME BEFORE DEPLETION

less allowable depletion (Figure 11-4)

DEPLETABLE ITEMS
Leasehold costs
Capitalized G&G
Lease Bonus
Capitalized IDC's (Figure 11-2)

TAXABLE INCOME

Taxable Income x Effective Tax Rate
Less: Investment Tax Credit
Plus: Adjustment Due to Minimum Tax Calculation
Equals: Cash Taxes

TABLE 11-1
Expenditures Incurred in Preparing A Well for Production

TRANSACTION	CLASSIFICATION
Cost of building a road to location to be used for drilling	IDC
Cost of preparing drill site including mud pits and drilling pad.	IDC
Cost incurred to negotiate drilling contracts.	IDC
Surface damage payments for damages during drilling	IDC
Cost of transporting rig to location	IDC
Cost of technical services required during drilling. (Engineers, geologists)	IDC
Cost of mud and fluids, drill stem tests	IDC
Cost of surface casing	Tangible
Cementing of surface casing	IDC
Labor to install casing	IDC
Cost of perforating casing	IDC
Completion costs—frac and acidize well	IDC
Swabbing costs	IDC
Costs to cleanup after drilling of well	IDC
Cost of tubing	Tangible
Cost of wellhead (Christmas Tree)	Tangible
Cost of Labor to install tubing	IDC
Cost of Labor to install Christmas tree	IDC

a statement is *not* made electing to expense IDC's, it is assumed that the taxpayer elects to capitalize IDC's. Thus a statement should be made with the first return filed, indicating the taxpayers election. The election, once made, is binding on all future transactions. Almost all taxpayers elect to expense IDC's because of the immediate tax savings. If IDC's made prior to 1983 are expensed, all expenditures classified as IDC's may be deducted from gross income in the period incurred if accural accounting is used, or in the period paid if on a cash basis. Subsequent to 1982, the enactment of the Tax Equity and Fiscal Responsibility Act of 1982 (TEFRA) reduces the amount of IDC's that a major can expense by 15%. The 1984 Tax Reform Act increased the amount to 20% of their "expensed" IDC's beginning in 1984. Starting in the month the expenditures are incurred, straight line depreciation is used to depreciate the 20% amount over 36 months.

If IDC's are capitalized, the intangible drilling costs are recovered through depletion and depreciation. Tangible items and the labor to install the tangible item (such a casing) are recovered through depreciation. Costs of drilling the well not represented by physical property are recovered through depletion.

The flow chart presented as Figure 11-2 shows the decision process for IDC expenditures.

11.5.2.7 *Depreciation*

Depreciation is a deduction in calculating taxable income that allows for the recovery of the cost of tangible items used in a trade or business. Prior to 1981, the principle of the depreciation deduction was to match revenues with the expenditures required to produce the income, thus following the matching principle used by accountants. However, the enactment of the Economic Recovery Act of 1981 and the establishment of the Accelerated Cost Recovery System (ACRS) changed this. Now, for most tangible items, the rate of recovery is not related to the productive life of the asset. Rather a system was established to recover, over a predetermined period not related to the useful life of the asset, the cost of most tangible assets. The cost recovery methods and periods are the same for both new and used property. Salvage value is assumed to be zero.

Under ACRS, tangible items are classified as 3, 5, 10, or 15 year property. The ACRS deduction is calculated by applying a statutory amount (found in appropriate tables) to the unadjusted basis (the cost) of the property. Since ACRS depreciation or alternatives to it must be used for most tangible depreciable property placed in service after 1980, only this method will be discussed. Other methods used to calculate depreciation for property placed in service prior to 1981 are presented in Appendix D.

Application of the ACRS method of depreciation requires two pieces of data: (1) the year the asset is placed in service; and (2) the class of property (3, 5, 10, or 15 year) the asset belongs. The classes of ACRS property are defined in terms of Section 1245 class property (tangible personal), Section 1250 class property (real), and the old Asset Depreciation Range (ADR) class life of assets in effect January 1, 1981. The ADR System established an acceptable depreciation period, or range of possible depreciation periods, for classes of property that could be used in making depreciation calculations.

11.5.2.7.1 *Classes of Property*

Only the 3-year and 5-year property classes will be discussed, since most of the tangible expenditures will fall into these two categories. Basically 5-year class property includes equipment such as pumping units, compressors, separators, and petroleum storage facilities. The 3-year property class includes automobiles and light duty trucks, and equipment used in connection with research and experimentation. It should be noted that some transactions in oil and gas may involve 10-year and 15-year property (depreciable real property such as buildings), and 15-year public utility property. As with all tax matters, the specific tax code requirements for a particular class of property should be checked.

11.5.2.7.2 *Calculating ACRS Depreciation*

An algorithm for calculating ACRS depreciation is presented after Figure 11-2. A flow chart

FIGURE 11-2

Recovery of Drilling and Completion Costs

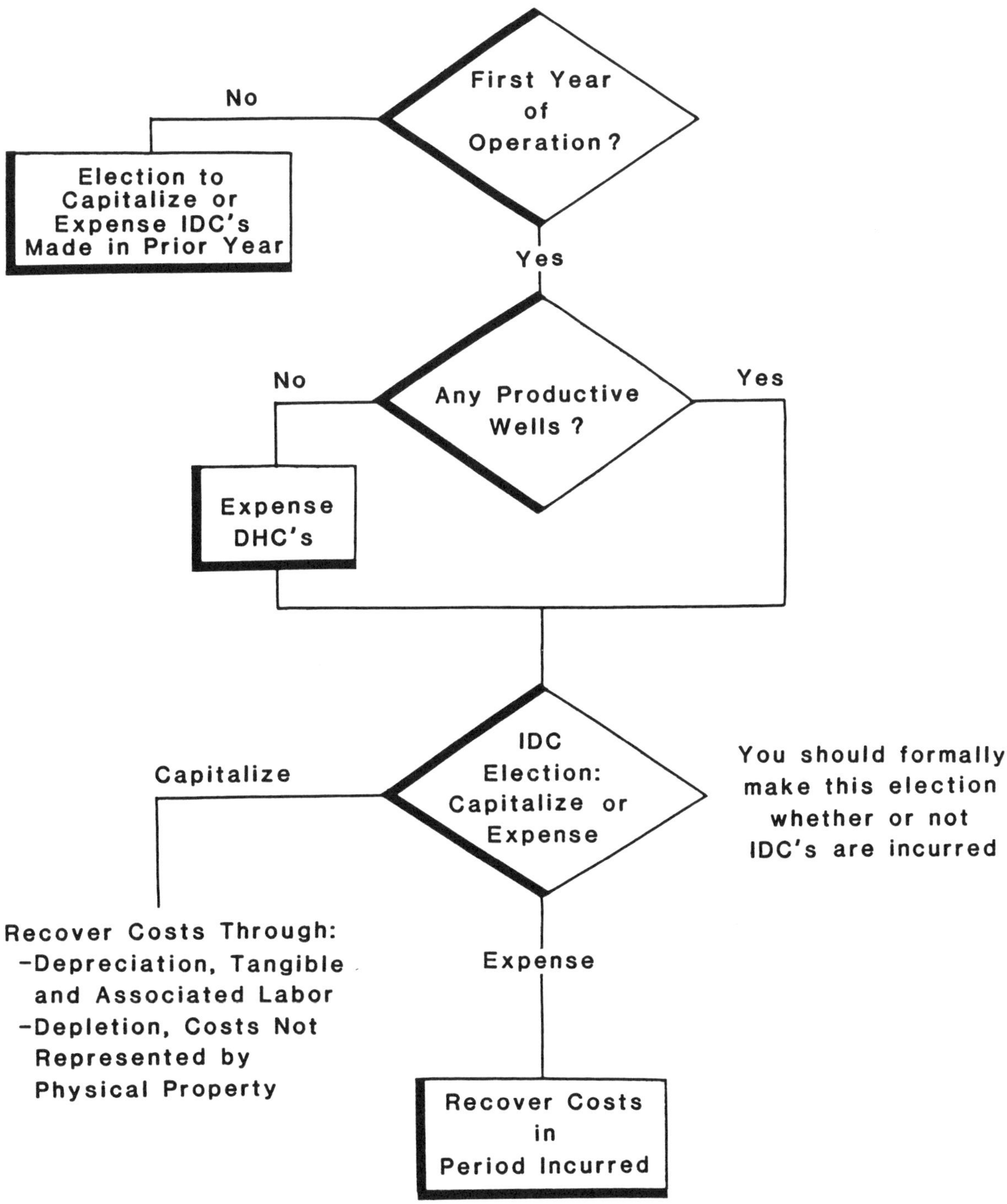

representing the recovery of tangible items is included as Figure 11-3. Depreciation rates for each class property and optional ACRS straight line rates are presented as Tables 11-2 and 11-3 respectively.

1) Determine the unadjusted basis for property placed in service. After 1982 the basis must be reduced by ½ of the investment tax credit (ITC) taken unless a specific reduced investment tax credit is used. ITC's will be discussed later in the chapter. The unadjusted basis is typically the cost of the property.

2) Determine the class of property (3, 5, 10, or 15). Select the appropriate statutory rate (Table 11-2) that is multiplied times the unadjusted basis for each year unless the election is made to use straight line depreciation. Straight line depreciation may be based on the recovery period of the property class (3, 5, 10, or 15 year) or optional extended periods as summarized in Table 11-3.

11.5.2.8 *Allowable Depletion*

IRC Section 611 provides that the economic interest owner of oil and gas reserves is allowed a deduction in calculating taxable income for the depletion of the natural resource. For oil and gas, allowable depletion is the greater of cost depletion or percentage depletion. Not all producers, however, are eligible for percentage depletion. The following paragraphs will demonstrate the cost depletion and percentage depletion calculations.

11.5.2.8.1 *Cost Depletion*

Cost depletion is calculated by determining the basis for the taxable year and dividing this amount by the number of units of reserves remaining at the beginning of the taxable year. The result is

Table 11-2
ACRS Depreciation Rates
Property Placed in Service After December 31, 1980

	Class of Investment			
Ownership Year	3-Year	5-Year	10-Year	15-Year Utility Property
	%	%	%	%
1	25	15	8	5
2	38	22	14	10
3	37	21	12	9
4		21	10	8
5		21	10	7
6			10	7
7			9	6
8			9	6
9			9	6
10			9	6
11				6
12				6
13				6
14				6
15				6
	100	100	100	100

then multiplied by the number of units sold during the taxable year. That is, what fraction of the total resource was depleted during the year, and accordingly what fraction of the remaining cost (adjusted basis) should be written off. In equation form, cost depletion can be represented as:

$$CD = AB\left(\frac{S}{U + S}\right) \qquad (11\text{-}1)$$

Where:

CD = Cost Depletion
AB = Adjusted basis for taxable year
S = Number of units sold during the year
U = Number of units of reserves at the end of the taxable year.

Reserves, adjusted basis, and the number of units sold are for the taxpayer's economic interest.

Items included in the depletable basis for cost depletion are geologic and geophysical costs, lease bonus paid by the lessee, and capitalized IDC's not represented by physical property.

11.5.2.8.2 *Percentage Depletion*

Percentage depletion is calculated on the basis of a specified percentage of the depletable gross income from the property not to exceed 50% of the taxable income before depletion from the property. Percentage depletion is a statutory concept and was significantly revised by the 1975 Tax Reduction Act and the 1976 Tax Reform Act. Percentage depletion was generally repealed by the 1975 Act. Certain exceptions to the act, however, do exist. Production from certain domestic gas wells and certain independent producers and royalty owners production is still eligible for percentage depletion. These two areas will be discussed in the following paragraphs.

11.5.2.8.2.1 *Certain Natural Gas*

The specific requirements for depletion on certain domestic gas production will not be covered. However, the general categories are: (1) regulated natural gas, (2) natural gas sold under a fixed contract, and (3) geothermal production. Specific requirements for gas in each category should be checked to determine the eligibility of the gas production. Table 11-5 presents the statutory depletion rates to be applied to the above categories of gas.

11.5.2.8.2.2 *IPRO Production*

The independent producers and royalty owners (IPRO) exemption from the Tax Reduction Act of 1975 allows percentage depletion on a limited amount of production by taxpayers who qualify as independents or royalty owners. An independent is a producer whose refinery runs do not exceed 50,000 barrels per day on any given day during the taxable year, and whose retail sales do not

Table 11-3
Optional ACRS Straight Line Depreciation

1. Straight Line Optional Recovery Periods for Personal Property

Class of Property	Possible Recovery Period
3 — Year	3, 5, or 12 years
5 — Year	5, 12, or 25 years
10 — Year	10, 25, or 35 years
15 — Year Public Utility	15, 35, or 45 years

NOTE: Half Year convention is followed in the first year the property is placed in service and the year following the recovery period.

2. Straight line Optional Recovery Periods for Real Property are 15, 35, and 45 years.

FIGURE 11-3

Recovery of Tangible Costs

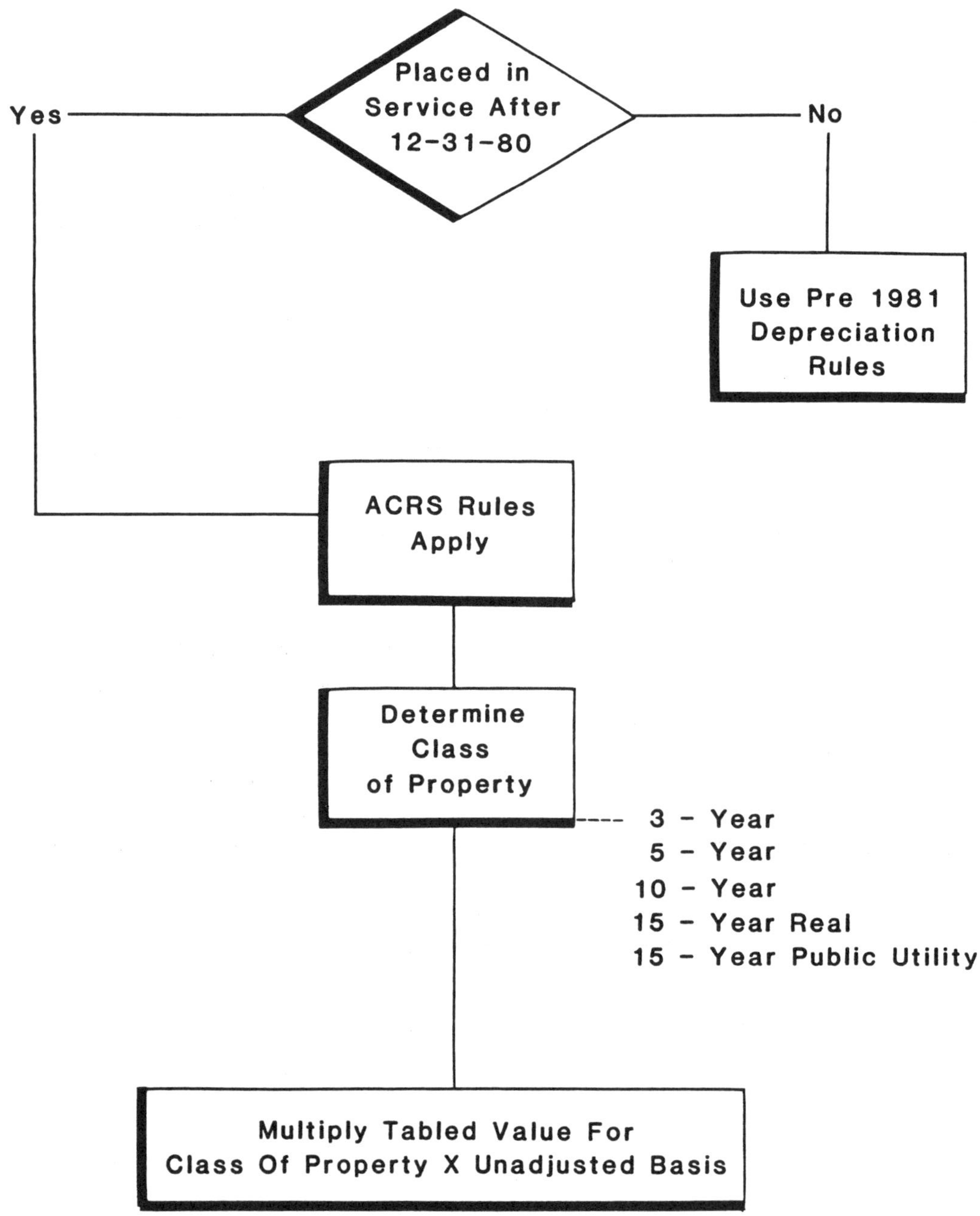

After 1982, basis is reduced by 1/2 of Investment Tax Credit (ITC) taken unless a specified reduced ITC rate is used.

exceed $5 million during the taxable year. The IPRO exemption is limited to the smaller of 50% of taxable income before depletion for the property or 65% of the taxpayer's taxable income before depletion, with certain adjustments, on an aggregate basis. Only a limited quantity of IPRO production is subject to percentage depletion. Table 11-4 presents the amount of IPRO production eligible for percentage depletion. Gas may, at the election of the taxpayer, be converted to it's oil equivalent at the rate of 6 MCF per barrel. Gas qualifying for percentage depletion such as regulated natural gas, fixed contract natural gas, or geothermal production, is not included in the following depletable amounts.

Table 11-4
Exempt IPRO Production
For Percentage Depletion

Year	Depletable Quantity (BOPD)
1975	2000
1976	1800
1977	1600
1978	1400
1979	1200
1980	1000
After 1980	1000

Additionally, the statutory rate of percentage depletion has been reduced from 22% in 1980 to 15% beginning in 1984 for primary production. Secondary and tertiary production, however, had a statutory depletion rate of 22% through 1983.

Effective January 1, 1984, the distinction between primary and secondary or tertiary oil is eliminated for purposes of percentage depletion. Independent producers and royalty owners can claim percentage depletion on all three types of qualifying production at a rate of 15%. Be careful with property transfers, however. *Generally*, beginning in 1984 percentage depletion is *not* available on secondary or tertiary production on *proven* production transferred after 1974. The Tax Reduction Act of 1975 had *generally* eliminated percentage depletion to the transferee for property transfers of proven *primary* production after 1974 but had exempted secondary and tertiary production.

11.5.2.8.3 *Depletion Allowance Calculation*

The following section will summarize the depletion allowance calculations. The special treatment of some items such as the lease bonus payments by the lessee are discussed here. Figure 11-4 presents a flow chart for the allowable depletion calculation.

1. Determine depletable gross income. Gross income calculated for our Federal Income Tax Model needs to be reduced by the amount of the lease bonus that has been depleted by the lessor. This assumes that the lease bonus is treated as an advanced royalty and is depletable by the lessor.

Table 11-5
Statutory Depletion Rates
for Certain Natural Gas Production
(%)

Year	Regulated Gas	Fixed Contract Gas	Geothermal Deposit
1981	22	22	22
1982	22	22	18
1983	22	22	16
1984	22	22	15
After 1984	22	22	15

Reducing the gross income in this manner is an application of the "bonus exhaustion" rule. The application of this rule is controversial and beyond the scope of this text.

2. Determine cost depletion. Cost depletion is calculated using equation 11-1. Depletable items include geologic and geophysical costs, lease bonus payments by the lessee, and capitalized IDC's not represented by physical property.

3. Determine percentage depletion. First, the production and taxpayer must meet certain specfic requirements to be eligible for percentage depletion. If these conditions are met:

   a) Multiply the appropriate statutory rate times the gross depletable income.

   b) Apply 50% net income limitation. Multiply the taxable income before depletion by 0.50. When calculating the taxable income before depletion, use gross income, not gross depletable income. For an IPRO, also apply the 65% net income limitation.

   c) Take the smaller of part a or b.

4. Allowable depletion is the greater of part 3 (c) or part 2.

5. Reduce the depletable basis (but not below zero) by the amount of the allowable depletion. The adjusted basis for the taxable year reduced by allowable depletion (but not below zero) is the adjusted basis for the next taxable year. Thus, using percentage depletion it is possible to recover more than your original depletable basis.

11.5.3 *Investment Tax Credit*

A tax credit, such as the investment tax credit (ITC), is subtracted from the computed tax liability and reduces the amount of tax due on a dollar for dollar basis. To qualify for the investment tax credit certain conditions must be met:

-- must be depreciable property with an ACRS recovery period of 3 or more years.

-- must be tangible personal property or real property used as an integral part of production.

-- must be used primarily in the United States.

-- must be placed in service during the taxable year.

Generally the tax rate for computing the ITC is 10%. However, 3 year ACRS property has an

Table 11-6
IPRO
Statutory Depletion Rate
(%)

Year	Primary Production	Secondary and Tertiary
1980	22	22
1981	20	22
1982	18	22
1983	16	22
after 1983	No distinction, all 15%	

Example 11-1
Intangible Drilling Costs

The following cost estimates for an authority for expenditure (AFE) to drill and complete a 3,900 foot well in Louisiana have been prepared.

		Tax Treatment	
	Cost	Dry Hole	Completion
Administrative Overhead	3,500	L	IDC
Location — Stake/Permit/Damages	2,000	L	IDC
Footage @ $10/Ft.	39,000	L	IDC
Company Labor — to drill	1,500	L	IDC
Cement & Cementing			
Surface casing	5,000	L	IDC
Production casing	10,000		IDC
Coring Service	1,500	L	IDC
Logging	12,000	L	IDC
Mudlogger	2,000	L	IDC
Engineering while drilling	5,000	L	IDC
Mud, water, fuel	10,000	L	IDC
Welding and Casing Crew			
Surface casing	1,500	L	IDC
Production casing	6,000		IDC
Completion Unit	14,000		IDC
Company Labor — to complete	5,000		IDC
Perforating	7,000		IDC
Casing — surface	15,000		T
— production	33,000		T
Tubing	15,000		T
Wellhead equipment	25,000		T
Float equipment, centralizers			
Surface casing	1,500		T
Production casing	3,000		T
Packers	5,000		T
Line pipe	1,500		T
Tanks & fittings	6,000		T
Separator & fittings	7,000		T

The column indicating the ''Tax Treatment'' shows which costs will be deducted as a loss (L) if the well is abandoned as a dry hole, and which costs would be treated as IDC's and tangibles (T) if the well is completed assuming the taxpayer has elected to expense IDC's. Usually the AFE will designate both the dry hole cost and the cost to drill and complete.

FIGURE 11-4

Depletion Allowance

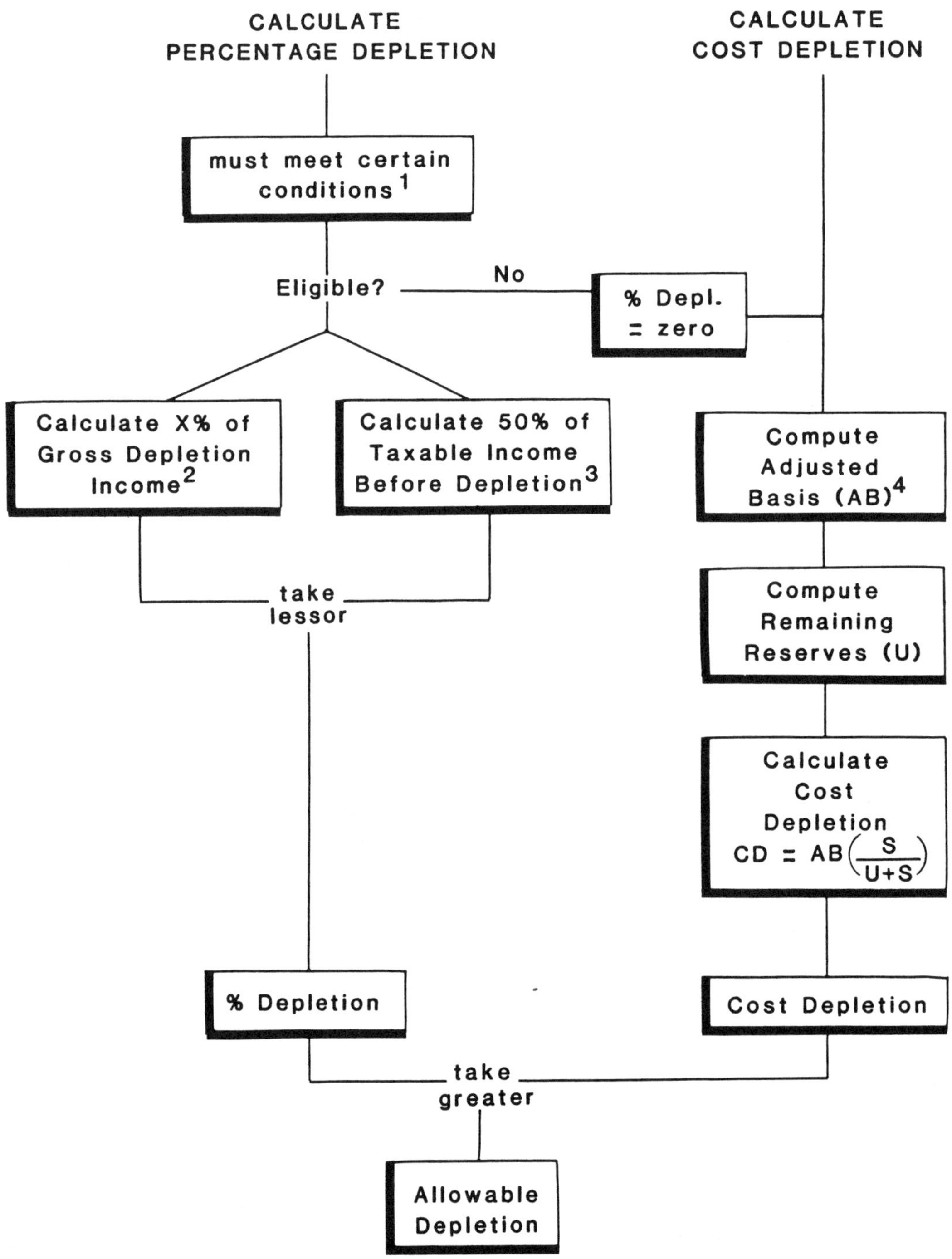

[1] See Notes on following page

ITC rate of 6%. Subsequent to 1982, either the investment tax credit is reduced by 2%, or the basis for ACRS depreciation is reduced by one-half of the ITC taken. Both new and used property are eligible for the investment tax credit. For used equipment only $125,000 counts in a given year. Any unused credit for used property cannot be carried forward or back. New property investment tax credits are limited to $25,000 of tax credit plus 90% of any remaining tax liability for 1982, and 85% of any remaining tax liability after 1982. In the case of tax credits on new property, any unused credit can be carried back three years and forward for 15 years.

If ACRS property is sold before the end of the recovery period, some of the ITC taken must be recaptured. Any ITC that is recaptured is added to the tax liability in the taxable year the recapture occurs. Table 11-7 presents the recapture schedule.

Examples of qualified investment property are:

Tangible personal property
- factory machinery
- office equipment
- trucks

Real Property — Integral part of production
- docks
- oil and gas pipelines
- oil derricks

Notes to
Figure 11-4

1. Percentage depletion is applicable to the following categories of domestic oil and gas:

   a. Some regulated natural gas.

   b. Natural gas sold under fixed price contract in effect before 1971.

   c. Geothermal Deposits.

   d. Independent producer and royalty owner (IPRO). After 1979 there is a 1000 BOPD limit on the amount of oil subject to percentage depletion.

2. Statutory Rates for the above classifications are as follows:

Rate (X%)

Classification

Year	a	b	c	d
1981	22	22	20	20
1982	22	22	18	18
1983	22	22	16	16
1984+	22	22	15	15

3. IPRO also has 65% of taxable income limitation on an aggregate basis.

4. AB = Prior period AB less allowable depletion from prior period.

Research and storage facilities
oil and gas storage tanks.

11.5.4 *Minimum Tax*

The minimum tax concept is intended to ensure that high income noncorporate taxpayers, through the use of items of tax preference, cannot avoid all taxes. The application of this concept is, unfortunately, very complicated. Our objective here will be to establish only the basis and concept for minimum tax calculations. Again, as with all taxation questions, a tax professional should be consulted concerning specific questions.

11.5.4.1 *Tax Preference Items*

Items of tax preference are presented in IRC Section 57. Only two tax preference items relating to oil and gas activities will be discussed here; depletion and intangible drilling costs.

11.5.4.1.1 *Depletion*

Percentage depletion allows for an ''allowable depletion'' deduction in excess of the basis of a property. To the extent that allowable depletion in a given taxable year exceeds the adjusted basis of the property, the excess is an item of tax preference. For example, if the adjusted basis for a property is $100,000 before any depletion deduction is made, then percentage depletion of $120,000 results in $20,000 of ''excess'' depletion. The $20,000 excess depletion is an item of tax preference. In subsequent years, all percentage depletion becomes ''excess'' depletion assum-

---

Example 11-2
Capital Recovery Using Accelerated Cost Recovery System

Using the AFE data presented in Example 11-1, calculate the ACRS depreciation for the property. The property qualifies as 5-year property and was placed in service during 1982.

The unadjusted basis for the tangible items is:

Casing — surface	$ 15,000
— production	33,000
Tubing	15,000
Wellhead	25,000
Float Equipment	4,500
Packers	5,000
Line Pipe	1,500
Tank & Fittings	6,000
Separator & Fittings	7,000
	$112,000

Using Table 11-2 to determine the ACRS recovery percentage

Year					ACRS Depreciation
1982	0.15	x	$112,000	=	$ 16,800
1983	0.22	x	112,000	=	24,640
1984	0.21	x	112,000	=	23,520
1985	0.21	x	112,000	=	23,520
1986	0.21	x	112,000	=	23,520
					$ 112,000

ing that the adjusted basis for depletion remains zero.

11.5.4.1.2 *Intangible Drilling Costs*

Intangible drilling costs are an item of tax preference to the extent that "excess IDC's" exceed "net income" from all oil and gas properties.

Obviously, one must know what is meant by "excess IDC's" and "net income". Burke and Bowhay (1983, p. 1107) have an example and discussion of this topic. The following algorithm is presented to calculate the amount of the tax preference.

1. Determine "Excess IDC's"

    IDC's on productive wells

    Less: Amortized amount for the year based on 120 months and straight line amortization. Amortization begins in the month of first production. Taxpayer may elect to use any method permitted for the cost depletion calculation such as units of production.

    Equals: Excess IDC.

2. Determine Net Income from ALL Oil and Gas Properties

    Gross Income

    Less: Allowable Deductions (see tax model)

    Plus: Deductions from non producing properties (DHC's)

    Plus: Excess IDC's

    Equals: Net Income

Example 11-3
Percentage Depletion

Assume that the gross depletable income in 1984 for a property is $100,000 and the taxable income before depletion is $40,000. The producer qualifies for the IPRO exemption. The statutory depletion rate is 15%.

A. If cost depletion is computed to be $1,000, what is the allowable depletion for the year?

B. If the adjusted basis prior to the allowable depletion is $10,000, what is the new adjusted basis?

A. Percentage depletion:

The lesser of:
0.15 * Gross Depletable Income
0.15 * 100,000 = 15,000

or

0.50 * Taxable Income before depletion
0.50 * $40,000 = 20,000

Therefore percentage depletion is limited to $15,000.

Allowable depletion:

The greater of:

Cost depletion or percentage depletion
Cost depletion = $1,000
Percentage depletion = $15,000

Therefore allowable depletion is $15,000.

B. The adjusted basis for the taxable year is reduced by the amount of allowable depletion but not below zero. For this case the adjusted basis becomes zero ($10,000 − $15,000 = negative).

---

3. Calculate the amount of the tax preference.

Excess IDC's

Less: Net income from all oil and gas properties

Equals: Amount of tax preference

11.5.4.2 *Noncorporate Taxpayers*

Subsequent to 1982, individuals are subject to an alternative minimum tax calculation. Prior to 1983 two calculations were necessary: an add-on minimum tax calculation and an alternative minimum tax calculation. Only the alternative minimum tax calculation subsequent to 1982 will be presented (IRC Section 55).

AMT =0.20 (AMTI—K)

where

AMT = alternative minimum tax
K = \$30,000 if a single taxpayer and \$40,000 for joint or surviving spouse returns
AMTI = alternative minimum taxable income

The alternative minimum taxable income is the adjusted gross income without regard to the net operating loss deduction (Code Sec. 172) *less* the

Example 11-4
Investment Tax Credit
and Recapture Calculations

Assume three-year ACRS property costing \$6,000 that qualifies for the ITC and is placed in service during 1983.

a) What is the ITC if the full amount of credit is taken?

Three year property qualifies for a full 6% tax credit.

\$6,000 x 0.06 = \$360

b) What is the basis for ACRS depreciation for 1983?

Basis = Cost − ½ ITC taken
= \$6,000 − \$180 = \$5,820

c) How much of the ITC from part a would be recaptured if the property is sold after one year but before two years?

From Table 11-5, 2/3 of the ITC taken must be recaptured.

⅔ x 360 = \$240 recaptured.

Table 11-7
Investment Tax Credit Recapture Schedule

Property Disposed (from date placed in service)	Fraction of ITC Recaptured 3-Year	5, 10, 15 — Year
within 1 year	1.0	1.0
After 1 year	2/3	0.8
After 2 years	1/3	0.6
After 3 years	0	0.4
After 4 years	0	0.2
After 5 years	0	0.0

In cases where the basis of the ACRS Property was reduced by ½ of the ITC, the basis must first be increased by 50% of the recaptured credit.

alternative net operating loss deduction, allowable itemized deductions, and accumulations distribution from a trust under Code Sec. 667, *plus* the sum of tax preferences.

The taxpayer then compares the amount of the alternative minimum tax (AMT) with the amount of "regular tax" for the year. The excess of the AMT over the regular tax amount is the amount added as an "adjustment due to minimum tax calculation" in the Federal Income Tax model (Exhibit 11-2) to arrive at the cash taxes.

11.5.4.3 *Corporate Taxpayers*

Corporations are subject only to the add-on minimum tax. The corporate minimum tax is covered in IRC Section 56. The add-on minimum tax is calculated using the following equation:

$$AOMT = 0.15\,(TPI - K)$$

where:

AOMT = add-on minimum tax
TPI = Tax preference items
K = the greater of \$10,000 or the full amount of the regular tax liability for the year.

Example 11-3
Cost Depletion

A taxpayer has a working interest in an oil property that is estimated to have approximately 40,000 barrels of oil net to his working interest measured at the end of the taxable year. The original capitalized leasehold costs are \$40,000. The adjusted basis for the taxable year is \$30,000. What is the cost depletion for the taxable year if 5,000 net barrels were sold during the year?

$$CD = AB\left(\frac{S}{S+U}\right)$$

AB = \$30,000
S = 5,000 bbl.
U = 40,000 bbl.

$$CD = (\$30,000)\left(\frac{5,000}{5,000 + 40,000}\right) = \$3,333$$

References

Burke, F.M., and Bowhay, R.W., 1983, Income Taxation of Natural Resources, PrenticeHall, Inc., New Jersey.

Federal Tax Guide, 1983, Commerce Clearing House, 4025 W. Peterson Ave., Chicago, Illinois.

Smith, C.L., Anderson, G.W., and Jones, M.A., 1982, Federal Taxation of Oil and Gas Transactions, Matthew Bender, New York.

Williams, H.R. and Meyers, C.J., 1981, Oil and Gas Law, Abridged Edition, Matthew Bender, New York.

PROBLEMS

11-1 For the following case assume:

- IDC's are expensed
- Depletable basis for year 1 is $10,000
- The producer is an IPRO and all production qualifies
- For percentage depletion use 16%, 15%, and 15% for years 1 through 3 respectively
- No investment tax credit or minimum tax calculations
- Tax savings can be used in the taxable year that the tax savings occurs

Year	Reserves End of Year	Production	Gross Income	Operating Costs	IDC
0					80,000
1	10,000	4,000	100,000	20,000	
2	5,000	600	20,000	20,000	
3	2,000	900	30,000	20,000	

a) Calculate the NCF's before FIT and draw the NCF diagram.
b) Calculate the NCF's after FIT and draw the cash flow diagram.
c) For parts a and b, assume year-end discounting to calculate the NPV discounted at a 10% annual rate.

11-2 The Big Buck Company, an independent producer, has incurred tangible expenditures of $200,000 (5 year ACRS) in *1981* for property "A". The depletable basis of property "A" is $380,000 at the beginning of 1982. All of Big Buck's production qualifies for percentage depletion. The following additional information is for 1982.

Sales = 17,000 STB
Reserves at beginning of 1982 = 50,000 STB
Operating costs (Big Buck's share) = $55,000
Production taxes (Big Buck's share) = $5,000
Oil Price = $30/STB
Big Buck's income tax rate is 46%

a. What is the cash income tax liability for *1982*?

b. What is the net cash flow for *1982*?

11-3 A new centrifugal pump is purchased and installed on January 1, 1982 for $12,000. The pump has been properly classified as a 5 year class property for purposes of the ACRS method of capital cost recovery. Your income tax rate is 48%. (Assume that any tax savings attributable to the purchase of the pump can be used in the year the tax saving is calculated.) Additionally, the pump qualifies for the 8% investment tax credit (the depreciable basis is therefore $12,000). On an after tax basis, what is the payout period (years) if the installation of the pump results in an annual operating cost savings of $1500. Assume the gross income is $50,000 per year either with or without the new pump. Assume operating costs are $5500 per year with the old pump and $4000 per year with the new pump.

11-4 Identify each of the following items as lease and well equipment (L&W), leasehold costs (LHC), lease operating expense (LOE), or intangible drilling costs (IDC).

Tubing
Casing
Labor to install casing
Completion costs
Perforating costs
Logging costs
Labor to install surface equipment
Pumping Unit
Lease bonus
Seismic and geologic expenditures
Wellhead

ADDENDUM
(2-10-89)

TAX CHANGES FOR CORPORATIONS
1986 TAX REFORM ACT

The 1986 Tax Reform Act made some changes in the tax law affecting the oil and gas industry. Some of these changes are summarized below:

INTANGIBLE DRILLING COSTS

INDEPENDENT - No change for Domestic wells. Must capitalize foreign IDC's. May recover costs over 10 year period using straight line amortization or may add to depletable basis.

INTEGRATED PRODUCER - May elect to expense 70% of IDC's in year incurred and amortize the remaining 30% over 60 months for domestic wells. Amortization begins in the month costs are incurred. Must capitalize foreign IDC's. May recover costs over 10 year period using straight line amortization or may add to depletable basis.

DEPRECIATION

The new depreciation is called Modified Accelerated Cost Recovery System (MACRS). Most oil field equipment is now classified as 7 year property. The method of depreciation is 200% declining balance switching to straight line. The asset is assumed to be placed in service in the middle of the year. Thus the depreciation schedule for 7 year property covers an eight year period. The salvage value is assumed to be zero. The following table summarizes the depreciation schedule for 7 year property.[1]

Year Placed in Service	Recovery
1	.1428
2	.2449
3	.1749
4	.1249
5	.0893
6	.0893
7	.0893
8	.0446

INVESTMENT TAX CREDIT

The Investment Tax Credit (ITC) was eliminated for assets placed in service after 12-31-85.

DEPLETION ALLOWANCE

The rules for percentage and cost depletion are unchanged. The independent producer and royalty owners are still eligible for 15% statutory depletion. Integrated producers are generally ineligible for percentage depletion.

TAX PREFERENCE ITEMS

The following is a list of tax preference items:

1. Percentage depletion to the extent it exceeds the basis of the property is an item of tax preference. No change was made in the 1986 Tax Reform Act.

2. IDC's were added as a tax preference item for Corporations. IDC's are an item of tax preference to the extent that "excess" IDC's exceed 65% of the net income from all oil an gas properties. Excess IDC's are the amount the IDC's taken in a given year exceeds the amount that would have been taken if the IDC's had been amortized over a 120 month period (straight line). Dry hole costs are not considered IDC's.

3. "Excess" depreciation was added to the list of tax preference items. "Excess" depreciation is the amount the depreciation taken exceeds the amount of depreciation that would be taken if 150% declining balance was used assuming the alternative depreciation life of 14 years.

4. "Excess Book Income" is a new item of tax preference added for corporations. This tax preference item requires the aggregate calculation of book income for the corporation. The tax preference amount is 50% of the excess of pre-tax book income over the alternative minimum taxable income before the book income preference for 1987 through 1989. Starting in 1990, the book income preference will be 75% of the excess of current earnings and profits (defined in the tax code) over taxable income.[2]

CAPITAL GAINS TREATMENT

All gains are taxed at the ordinary income tax rates.

ALTERNATIVE MINIMUM TAX

The alternative minimum tax is calculated by taking the regular taxable income less certain deductions plus items of tax preference. The minimum tax rate of 20% is multiplied times the alternative minimum taxable income (AMTI) reduced by an exemption amount. The exemption amount is $40,000 less 25% of the AMTI over $150,000. The tax payer is liable for the larger of the regular tax liability or the minimum tax liability. When the AMT is larger that the regular tax, the difference is referred to as the "net minimum tax" and is really the additional tax due because of the minimum tax calculation.

A new feature is the minimum tax credit. When additional tax is paid as a result of the AMT calculation, this amount can be used to offset the amount the regular tax exceeds the AMT in future years. Thus the impact of the alternative minimum tax may only be temporary in that additional minimum taxes paid in early years may be recouped in subsequent years.

CORPORATE TAX RATES
1986 Tax Reform Act

Taxable Income	Rate
Not over $50,000	15%
Over $50,000, less than $75,000	25%
Over $75,000, less than $100,000	34%
Over $100,000, less than $335,000	39%
Over $335,000	34%

REFERENCES:

1. Prentice-Hall's Explanation of the Tax Reform Act of 1986, Prentice-Hall Inc., Paramus, N.J. (1986).

2. S. B. Snook and W. J. Magnuson, Jr.: "The Tax Reform Act's Hidden Impact on Oil and Gas," The Tax Adviser (Dec. 1986) pp 777-783.

3. Robert S. Thompson: Society of Petroleum Engineers, Paper No. 16309, The Impact of the New Tax Law on Internal Cash flow Generation, presented at the SPE Hydrocarbon Economics and Evaluation Symposium, Dallas, Texas, (March 2-3, 1987).

4. J. D. Wright, R. S. Thompson and S. E. Shipley: Society of Petroleum Engineers, Paper No. 16312, The Impact of the New TAx Law on Project Evaluation,presented at the SPE Hydrocarbon Economics and Evaluation Symposium, Dallas, Texas, (March 2-3, 1987).

Table of Contents
Chapter 12
The Report

12.1 **Introduction** 12-1
12.2 **Conclusions** 12-1
12.3 **Recommendations** 12-2
12.4 **Discussion** 12-2
12.5 **Appendix** 12-3
12.6 **Other Considerations** 12-3

Chapter 12

The Report

The best engineer in the world will fail if he cannot express his ideas both verbally and in writing. Ideas must be communicated to be of benefit to society. Whether a report is written to management or to a bank for the purpose of a loan, the report should be organized so that one can read a limited amount of the report and understand the reason for the report, the important conclusions, and the recommendations without having to spend an inordinate amount of time. Obviously, clarity is important since a poorly written report wastes time. There should be no doubt as to what the author is attempting to state. Remember, that "fuzzy" writing usually means "fuzzy" thinking.

Although there is no single format that is acceptable in all cases, the approach here is to present a suggested format for the oil property evaluation report. The proposed format accomplishes two things: (1) it minimizes management's time in getting the "meat" out of the report, and (2) it presents the important items of a rigorous oil evaluation report. This outline can be used as a checklist to insure that the important aspects of the report have been documented. Good documentation is important. Things that are perfectly clear today will be forgotten next year when the evaluation is updated. The following paragraphs will discuss the five sections of the suggested oil property evaluation report: introduction, conclusions, recommendations, discussion and appendices.

12.1 INTRODUCTION

The introduction should tell the uninformed reader the reason for the report and the objective of the report. A typical reason for a report might be to evaluate Joe Oilman's interest in the Big Deal Field located in some county of Oil Country. The objective of the report may be to satisfy the Securities Exchange Commission reporting requirements or to determine a bid price for the property. Usually, the introduction includes important location data for the study such as the lease name, field name, the county, and the state. Often the number of wells in this study are stated. In general, the introduction should tell the reader enough so that the conclusions and recommendations make sense. Experienced management will often know immediately if the conclusions and recommendations are reasonable because of their own experience in an area, or from just "what is reasonable" based on their past experiences. If there is any doubt, rest assured that the reader will look at more than just the introduction, conclusions, and recommendations.

12.2 CONCLUSIONS

The conclusions may be written in paragraph form. However, a numerical listing of the important conclusions is often preferred. The conclusions should be written in order of importance with the most important being first. The numerical approach is easy to read and understand. For an oil property evaluation, the conclusions probably would incorporate the following:

(1) Present value of the future production discounted to the effective date of the study at one or more interest rate(s). State if calculations are before or after federal income taxes.

(2) Remaining reserves as of the effective date of the study. Both gross and net

numbers should be stated. Separate summaries should be given for each type of economic interest evaluated (working interest, royalty, overriding royalty, net profits). Also separate summaries should be presented for proved developed producing, proved developed nonproducing, and proved undeveloped reserves.

(3) If a few wells dominate the total net remaining reserves, this is an important conclusion for management to be aware of.

(4) EUR and CUM summaries, both gross and net values.

12.3 RECOMMENDATIONS

Again, the recommendations should probably be listed numerically rather than in paragraph form. Examples of the type of statements that should be included in the recommendations are:

(1) Sell or buy recommendations. Give a summary of the basis for your buy or sell recommendation. For example, the bid is based on the PV discounted at X%.

(2) Field gas compression capacity should be increased by installing additional compressors costing ...

(3) The proposed waterflood should be implemented in two phases. This will increase reserves by ...

12.4 DISCUSSION

The discussion should defend all your conclusions and recommendations in addition to presenting detailed documentation for the entire study. As such, this is the section of the report that will be read if management is curious or is not convinced that your conclusions are valid. Accordingly, a rigorous analysis of the problem should be clearly presented. For an oil property evaluation the following items should be covered.

(1) Geologic data including cross sections, structure maps, gross and net pay maps for each formation evaluated, depositional environment.

(2) Reservoir Description
- a) Areal map
- b) Location of producing wells, dry holes
- c) PVT data (Rs, Bo, Bg, separator tests)
- d) Reservoir energy (solution gas, gas expansion, gravity drainage, water drive, combination)
- e) Average porosity, water saturation

(3) Reserves
- a) Analogy
- b) Volumetrics
- c) Decline Curves
- d) Material Balance

When data is available, each possible method of calculating reserves should be employed. The engineer's challenge is to explain "why" the various methods agree or, more likely, do not agree. Each method has certain assumptions and limitations. The equations, assumptions, and methods employed should be clearly stated. Any weakness in your calculations because of data should be clearly stated. It is useful to show your detailed calculations in an appendix. Suggested items to include in the appendix will be presented separately.

(4) Production Forecast

The evaluation is based on your reserve estimate and the associated production forecast.

- a) Production data should be plotted. Wild fluctuations should be explained.
- b) Projected production should be plotted as a dashed line giving management a visual check of the reasonableness of the forecast.
- c) Has downtime been taken into consideration?
- d) Method employed to project production.

(5) Economic Assumptions
- a) Effective date of study.
- b) Working interest and net revenue interest assumptions.

c) Oil and gas price assumptions including price escalations.
d) Cost escalation assumptions.
e) Windfall profit tax assumptions.
f) State and local tax assumptions.
g) Federal tax assumptions, if necessary.

Economic summary calculations for each well and for categories of production such as proved developed-producing, proved developed-nonproducing, and proved-developed should be included. The calculations for each well should be included in the appendix.

12.5 APPENDIX

Voluminous, detailed calculations can often be best documented in an appendix. Examples of some items that might best be presented in an appendix are:

(1) Summary log analysis sheets.
(2) Summary volumetric calculations for each well.
(3) Material balance calculations for the field.
(4) Production plots and forecast for each well.
(5) Summary of decline curve analysis for each well.
(6) Yearly net cash flow calculations for each well

12.6 OTHER CONSIDERATIONS

Depending upon the length and format of the report, a cover page, table of contents, list of figures, and list of tables may be appropriate.

All maps should contain:

(1) A north arrow showing magnetic declination if known.
(2) A key to symbols used.
(3) A graphic (bar) scale. The current, popularity of reducing and enlarging "Xerox" machines renders a numeric (1″ = 2000′) scale useless.
4) A title block giving:
   a. Kind of map (structure on E-marker, Isopach of Gross Dakota Interval, etc.).
   b. Name of field or prospect.
   c. Location (County or Parish, State).
   d. Bar Scale.
   e. Contour Interval.
   f. Names or Initials of Author and Draftsman.
   g. Date.

Each manager or reader of a report has a few "pet peeves." An author should attempt to avoid using words which will "set off" the reader. An example is beginning the Introduction with "The purpose of this report is . . .". A more imaginative lead-in is suggested. Perhaps, "The Babcock Field, Sweetwater County, Colorado, produces approximately 12,000 BOPD from the Niobrara Shale." An approach which might help smooth the report is to assume that you are going to read the Introduction, Conclusions, and Recommendations to the client or manager. You have no more than 5 minutes to present your case and will not be allowed to expand any points. You should ask yourself:

> Have I given the manager enough information to understand the problem?
>
> Is the story presented in a logical order?
>
> Have I presented all of the important conclusions?
>
> Can I read the report out loud (loudly — not whispering) and not stumble over sudden turns in logic or style?

If the answer to all of the questions is "yes," then the report probably communicates its message. The engineering? — now that's another question.

Appendix A

Derivation of Geometric Gradient Formula

In exponentially declining wells, the flow of funds decreases at a geometric rate. That is, the receipts for period j + 1 are equal to some constant times the receipts for period j. In equation form

$$A_{j+1} = R * A_j \qquad \text{(A-1)}$$

When product prices are constant and all costs can be expressed on a per unit of production basis, the net revenue from a property is proportional to the producing rate. The ratio between successive period's income for exponential decline is then:

$$R = e^{-D} \qquad \text{(A-2)}$$

where the units on D are 1/period. (1/year for yearly projections; 1/month for monthly projections).

Of course, other values of R are possible. If revenue is expected to double every period then R = 2.

The present value of the geometrically increasing or decreasing series can be expressed as

$$P = \frac{A_1}{(1+i)^1} + \frac{A_2}{(1+i)^2} + \frac{A_3}{(1+i)^3} + \ldots$$

$$+ \frac{A_n}{(1+i)^n} \qquad \text{(A-3)}$$

for end-period discounting

Since $A_{j+1} = R * A_j$

$$A_2 = R*A_1$$

$$A_3 = R*A_2 = R^2A_1$$

$$\vdots$$

$$A_n = R^{n-1} A_1$$

Substituting into (A-3)

$$P = A_1(1+i)^{-1} + RA_1(1+i)^{-2}$$

$$+ R^2A_1(1+i)^{-3} + \ldots$$

$$+R^{n-2}A_1(1+i)^{-(n-1)}$$

$$+ R^{n-1}A_1(1+i)^{-n} \qquad \text{(A-4)}$$

factoring out A_1 and multiplying by $(1+i)^{n+1}$

$$P(1+i)^{n+1} = A_1[(1+i)^n+R(1+i)^{n-1}$$

$$+ \quad R^2(1+i)^{n-2} \quad + \quad \ldots$$

$$+ R^{n-2}(1+i)^2 + R^{n-1}(1+i)] \qquad \text{(A-5)}$$

multiplying (A-4) by $R(1+i)^n$

$$PR(1+i)^n = A_1[R(1+i)^{n-1}+R^2(1+i)^{n-2}$$

$$+\ldots+ R^{n-2}(1+i)^2$$

$$+ R^{n-1}(1+i) + R^n] \qquad \text{(A-6)}$$

subtracting A-6 from A-5

$$P(1+i)^{n+1} - PR(1+i)^n$$
$$= A_1[(1+i)^n - R^n] \quad \text{(A-7)}$$

dividing by $(1+i)^n$

$$P(1+i) - PR = \frac{A_1[(1+i)^n - R^n]}{(1+i)^n} \quad \text{(A-8)}$$

factoring out P and dividing by $(1+i-R)$

$$P = A_1 \left[\frac{(1+i)^n - R^n}{(1+i)^n (1+i-R)}\right] \quad \text{(A-9)}$$

When $R = 1$ Equation (A-9) reduces to the uniform annual series equation (2-11).

For an oil property, A_1 equals the oil price less the operating costs times the first year production.

Example A-1
Present Worth of
Exponentially Declining Production

A well is producing at 100 BOPD (net to the interest being evaluated) and is expected to decline at 50%/year (nominal). If the economic limit is 3 BOPD and the margin (selling price less operating costs) is $20.00 per bbl, what is the present worth of the future production discounted at 10%?

Time to the economic limit

$$t = \frac{\ln\left(\frac{q_{el}}{q_i}\right)}{-D}$$

$$t = \frac{\ln\left(\frac{3}{100}\right)}{-.5}$$

$t = 7$ years

First year reserves

$$N_P = \frac{q_i\,(1 - e^{-D})}{D}\ 365$$

$$N_P = \frac{100\,(1 - e^{-.5})}{.5}\ 365$$

$N_P = 28{,}723$ STB

Ratio of 2nd year reserves to 1st year reserves

$R = e^{-D}$

$R = e^{-.5}$

$R = .6065$

Present value of production

$A_1 = 28{,}723$ STB * \$20/STB

$i = .10$

$n = 7$ years

$R = .6065$

$$P = A_1 \left[\frac{(1+i)^n - R^n}{(1+i)^n (1+i\text{-}R)}\right] \quad \text{(A-9)}$$

$$P = 574{,}460 \left[\frac{1.1^7 - .6065^7}{1.1^7\,(1.1 - .6065)}\right]$$

$P = 574{,}460\ [1.9950]$

$P = \$1{,}146{,}000.$

Appendix B

The Use of Loss Ratio to Extrapolate Decline Curves.

In 1927, Johnson and Bollens introduced the concept of ''loss-ratio'' to extrapolate decline curves. Loss ratio is defined as the production rate at the end of a period divided by the loss in production rate during the period. In 1935 Pirson showed that, for exponential decline, the loss ratio for all periods is constant. In hyperbolic decline, the difference between successive loss ratios is constant. Arps, in 1944, gave a review of the state of the art in decline curve extrapolation and published examples of the use of loss ratio to extrapolate both an exponential and a hyperbolic decline curve. These two examples have become the classic examples of decline curve extrapolation and have been referenced by many authors.

The method used by Arps to extrapolate a hyperbolic decline curve consists of calculating the first derivative of loss ratio for each period, averaging those derivatives, and then using that average to ''back calculate'' future production rates. Ramsay, in his masters thesis, proves that the method used by Arps requires only the first two production rates and the last two production rates. All production rates in between these four values cancel out of the equation and are therefore ignored! The equation for b that Ramsay derives is

$$b_{avg} = \frac{1}{n-2}\left(\frac{q_n}{q_{n-1} - q_n} - \frac{q_2}{q_1 - q_2}\right) \quad \text{(B-1)}$$

where

n = the number of values of production

q_j = the production rate during month j

Ramsay, in demonstrating the use of Equation B-1 uses Arp's example of hyperbolic decline. Ramsay calculates a value for b of .51 compared to Arps .508. Since the value of b calculated by both methods should be exactly the same, Arp's Table 2 was recalculated by the authors and a running average of the value of b was compared to the value calculated by Equation 1. Those calculations are attached as Table I. It was found that Arps original table had a number of relatively insignificant math errors and one significant math error. When the table was recalculated, it was found that Ramsay's equation and Arp's averaging technique gave *exactly* the same answer for all periods. Ramsay further presents an example in which the method of Arps predicts an increasing production rate rather than a decreasing rate. This occurs because the last period's production is greater than the next to last periods production. This example is reproduced in Table II. Notice that even though it is obvious from the data that the field is declining, the loss ratio method predicts an increasing producion rate.

Conclusions

Although the loss ratio method has been used for a number of years, Ramsay has shown that it depends solely on the first two period's production and the last two period's production. He has further shown that a single anomolous data point can result in meaningless answers. Because of these failings, the loss ratio method should never be used to extrapolate hyperbolic decline curves.

TABLE I

Month	Year	Monthly Prod Rate q	Loss in Prod Rate during 6 Months Interval Δq	Loss Ratio on Monthly Basis $a=6\frac{q}{\Delta q}$	First Derivative of Loss Ratio $b=\frac{\Delta(\frac{6q}{\Delta q})}{6}$	Running Average of b	b from Ramsay's Equation
Jan	37	28,200					
July		15,680	−12,520	− 7.5144			
Jan	38	9,700	− 5,980	− 9.7324	−.3697	−.3697	−.3697
July		6,635	− 3,065	−12.9886	−.5427	−.4383	−.4383
Jan	39	4,775	− 1,860	−15.4032	−.4024	−.4383	−.4383
July		3,628	− 1,147	−18.9782	−.5958	−.4777	−.4777
Jan	40	2,850	− 778	−21.9794	−.5002	−.4822	−.4822
July		2,300	− 550	−25.0909	−.5186	−.4882	−.4882
July	41	1,905	− 395	−28.9367	−.6410	−.5101	−.5101
July		1,610	− 295	−32.7458	−.6348	−.5257	−.5257
Jan	42	1,365	− 245	− 33.4286*	−.1138	−.4799	−.4799
July		1,177	− 188	−37.5638	−.6892	−.5008	−.5008
Jan	43	1,027	− 150	−41.0800	−.5860	−.5086	−.5086

* Arps Table 2 reports "a" as 34.43, and "b" as −.28. This error affects all of the remaining values of b in his table.

TABLE II

The following table shows an extrapolation of actual production data for a Kansas lease producing from the Lansing limestone using the method of average differential of loss ratio: (Data from Ramsay)

Year	Annual Production	Annual Loss in Production	Loss Ratio	First Derivative Of Loss Ratio
	q	Δq	a	Δa
1950	65,441	–	–	–
1951	55,509	−9,932	− 5.59	
1952	45,529	−9,980	− 4.61	+0.98
1953	35,642	−9,887	− 3.60	+1.01
1954	30,127	−5,515	− 5.45	−1.85
1955	27,876	−2,251	−12.40	−6.95
1956	23,492	−4,284	− 5.35	+7.05
1957	30,745	+7,253	+ 4.23	+9.58
1958	34,655	+3,910	+ 6.87	+1.64
1959	38,305	+3,650	+ 8.51	+1.64
1960	41,740	+3,435	+10.15	+1.64
1961	45,000	+3,260	+11.79	+1.64
1962	48,115	+3,115	+13.43	+1.64
1963	51,115	+3,000	+15.07	+1.64
1964	54,005	+2,890	+16.71	+1.64
1965	56,715	+2,710	+18.35	+1.64

Production data through 1957 is actual lease production. Tabulated values from 1958 through 1965 were computed from the average first derivative of loss ratio (equal to +1.64). Figure I compares the computed production values to the actual production for the period, 1958 to 1965. It is obvious from this example that the loss ratio method will give incorrect results when the production rate for the last year is greater than the rate for the next to last year.

References

Arps, J. J.: "Decline Curve Analysis," Trans AIME, 1945, Vol. 160, p. 228.

Johnson, R. H. and Bollius, A. L.: "The Loss Ratio Method of Extrapolating Oil Well Decline Curves," Petroleum Development and Technology, 1927, AIME Vol. 77, p. 771.

Pirson, S. J.: "Production Decline Curves of Oil Well May Be Extrapolated by Loss Ratio," Oil and Gas Journal, No. 26, 1935, P. 34.

Ramsay, H. J., Jr: "The Ability of Rate-Time Decline Curves to Predict Future Production Rates," M. S. Thesis, Univ. of Tulsa, 1968.

FIGURE

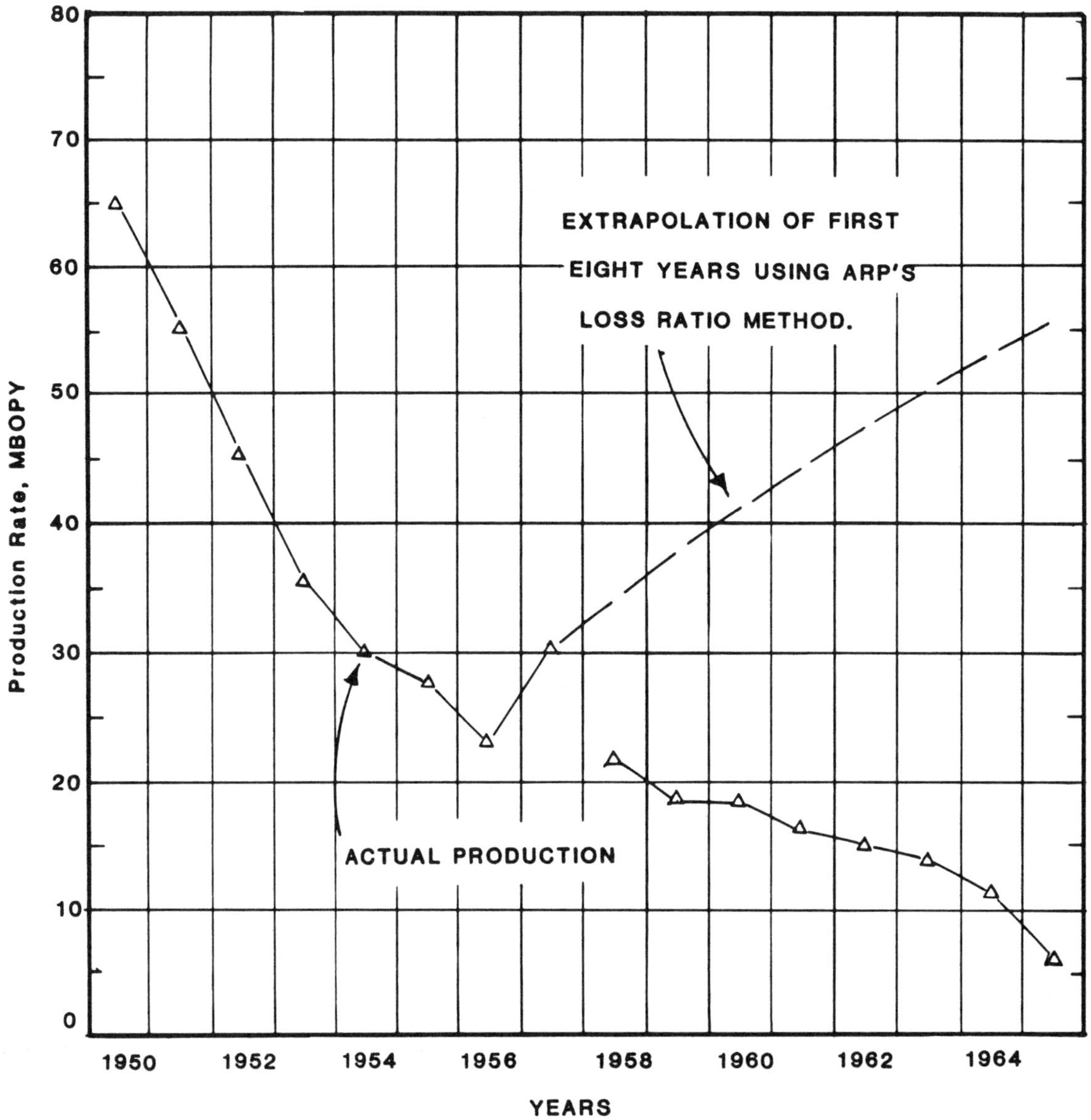
80
70
60
50
40
30
20
10
0
Production Rate, MBOPY
1950
1952
1954
1956
1958
1960
1962
1964
YEARS
EXTRAPOLATION OF FIRST
EIGHT YEARS USING ARP'S
LOSS RATIO METHOD.
ACTUAL PRODUCTION

Appendix C

Construction of Specially Scaled Graph Paper.

If we denote f(Z) as the function of Z which we intend to plot on the y—axis (for example $f(Z) = Z^{-1/m}$) then the distance in inches (for example) from the origin to any value of Z is:

$$Y = [\,f(Z) - f(Zmin)]\,SF$$

where

$$SF = \frac{\text{Length of Y-axis}}{f(Z_{max}) - f(Z_{min})}$$

This procedure can also be used for the X-axis.

Example C-1
Construction of two-cycle semi-log paper.
K+E #46-4973

$Z_{min} = 1$

$Z_{max} = 100$

Length of axis = 10 inches

$$SF = \frac{10}{\log_{10}(100) - \log_{10}(1)} = \frac{10}{2 - 0} = 5''$$

$$Y = [\log_{10}(Z) - \log_{10}(1)] * 5 = 5 \log_{10}(Z)$$

z	Y (inches)
1	0
5	3.495
10	5.000
20	6.505
50	8.495
100	10.000

The above values can be checked by measuring a piece of semilog paper.

Appendix D

Depreciation Methods Prior to 1981

Prior to 1981, four basic methods of depreciation were accepted by the tax code: (1) the straight line method, (2) the declining balance method, (3) the sum-of-the-year's digits method, and (4) the units of production method.

D.1 STRAIGHT LINE METHOD

The annual straight line depreciation expense is calculated by dividing the cost of an asset, reduced by the salvage value, by the life in years. One divided by the assumed life of the asset (in years) multiplied by 100 is the straight line depreciation rate in percent.

Example D-1
Straight Line Depreciation

A machine costing $11,000 is estimated to have an estimated salvage value of $1,000 at the end of 10 years. What is the annual depreciation expense?

For each year

$$(\text{Cost} - \text{Salvage Value}) * \frac{1}{10}$$

$$= (11{,}000 - 1{,}000) * \frac{1}{10}$$

$$= \$1{,}000 \text{ per year}$$

The straight line rate of depreciation is 10% per year.

D.2 DECLINING-BALANCE METHOD

The declining balance method is an accelerated method of depreciation. That is, the greatest deduction occurs in the first year. In each subsequent year, the deduction becomes smaller. The concept of declining balance depreciation can be applied to various rates of declining depreciation. Common rates of declining balance depreciation are: (1) 200 percent (double) declining balance, (2) 150 percent (1 ½) declining balance, and (3) 125 percent (1 ¼) declining balance.

The annual rate used to compute the depreciation expense is calculated by multiplying the computed staight line rate for the assumed life by the method of declining balance.

For example:

200 percent declining balance:

2 x straight line rate = annual rate for double declining balance

150 percent declining balance:

1.5 x straight line rate = annual rate for 150% declining balance

125 percent declining balance:

1.25 x straight line rate = annual rate for 125% declining balance

This annual rate is then multiplied times the remaining balance at the beginning of the taxable year to calculate the depreciation for the taxable year. Each year, the depreciable basis (remaining balance) is reduced by the depreciation deduction. Salvage value is ignored when calculating the depreciable basis. Although the salvage value is ignored in calculating the depreciable basis, the basis cannot go below a reasonable salvage value. The taxpayer may change from the accelerated rate of decline to straight line depreciation at any time. When the switch to straight line rate is made, consider only the number of years remaining in the life of the asset at the time of the switch.

Example D-2
Declining Balance Depreciation

Assume double declining balance depreciation for the asset in Example D-1.

Double declining balance rate = 2 x straight line rate
= 2 x 10%/year
= 20%/year
Ignore salvage value.

Year	Remaining Balance (End of Year)	Depreciation Expense
0	$11,000	$ 0
1	8,800	2,200
2	7,040	1,760
3	5,632	1,408
4	4,506	1,126
5	3,605	901
6	2,884	721
7	2,307	577
8	1,846	461
9	1,477	369
10	1,182	295

For example, the depreciation in year one is 0.20 x $11,000 = $2,200

The remaining balance or depreciable basis for year two is $11,000 − $2,200 = $8,800.

D.3 UNITS OF PRODUCTION

The units of production depreciation is the concept applied when cost depletion is calculated. The concept is that the useful life is not based on a specified period of time, rather the asset's usefulness is based on the number of units it is estimated to produce. The depreciation of the asset for the taxable year is then based on the fraction of the estimated remaining units that it actually produced during the year times the remaining undepreciated balance at the beginning of the year. In equation form:

$$UOP = AB \left(\frac{P}{P + R} \right)$$

where:

UOP = units of production depreciation
AB = adjusted basis for the taxable year. The adjusted basis for the subsequent year is calculated by subtracting the UOP depreciation for the taxable year from the beginning undepreciated balance.
P = actual production during the year.
R = estimated remaining production capacity for the asset at the *end* of the taxable year.

The original basis of the asset is decreased by the estimated salvage value when calculating the adjusted basis for the first year.

Example D-4
Units of Production Depreciation

Calculate the UOP depreciation for the asset in Example D-1 using the following assumed production schedule.

Year	Production (Units)	UOP Depreciation ($)
1	1,000	1,000
2	2,000	2,000
3	4,000	4,000
4	3,000	3,000

For year 1:

$$\left(\frac{1,000}{10,000} \right) (11,000 - 1,000) = 1,000$$

For year 2:

$$\frac{2,000}{9,000} (10,000 - 1,000) = 2,000$$

D.4 SUM-OF-THE-YEAR'S DIGITS METHOD

The sum-of-the-year's digit (SYD) method is also an accelerated technique of depreciation and can be expressed by the following equation.

$$SYD = \frac{RY}{\frac{n\,(n+1)}{2}}\,(C - SV)$$

where:

SYD = sum-of-year's digits depreciation expense for the year.
RY = Remaining number of years in the life of the asset.
n = the assumed life of assets (years)
C = cost of the asset
SV = salvage value of the asset

Example D-3
Sum-of-the-year's digits depreciation

Assume the asset purchased in Example D-1.

Sum of the years is:

$$\frac{n\,(n+1)}{2} = \frac{10\,(11)}{2} = 55$$

For Year 1:

$$SYD = \frac{10}{55} \times (11{,}000 - 1{,}000) = \$1{,}818$$

For Year 2:

$$SYD = \frac{9}{55} \times (11{,}000 - 1{,}000) = \$1{,}636$$

Year	SYD Expenses ($)
1	1,818
2	1,636
3	1,455
4	1,273
5	1,091
6	909
7	727
8	545
9	364
10	182
	$10,000

Index

Accelerated Cost Recovery System see ACRS
Acceleration projects 3-15, 3-17
Accounting procedures, operating agreements 7-6
Accounting rate-of-return 3-3, 3-5
Accrual 1-4
ACRS 1-9, 11-10, 11-12, 11-16
Adjusted base price 10-1, 10-2, 10-3
Advalorem taxes 9-1, 9-6
After casing point and before payout, trigger points 7-11, 7-15
After payout, trigger points 7-11, 7-15, 7-16
Allocated overhead, federal income taxation 11-7
Allowable deductions, federal income taxation 11-2, 11-6
Amortization
  defined 1-6
  financial model 1-4
Analogy 4-6, 4-13
Annuity (loan) 2-5
Annuity due or lease 2-5
AOEROR, appreciation-of-equity rate-of-return 3-21
API gravity 6-1
Appreciation-of-equity rate-of-return 3-21
Arps, author 4-2, 4-3, 5-37
Attributes of ideal yardsticks 3-2
Authority for Expenditure 8-3
Authors
  Arps 4-2, 4-3, 5-37
  Beal 5-1
  Cox 5-12, 5-13
  Cutler 5-37, 5-38
  Earlougher 5-13
  Ehlig-Economides 5-12, 5-14
  Fetkovich 5-12, 5-19, 5-27, 5-39, 5-40
  Garb 5-43
  Gentry 5-34
  Lefkovits 5-11
  Lewis 5-1
  Matthews 5-11
  Ramsay 5-37, 5-38
Back-ins 7-4
Bailout factor 3-3, 3-4
Barrels per acre foot, reserves 4-6
Base price, windfall profit tax 10-2
Beal, author 5-1
Before casing point, trigger points 7-11, 7-13, 7-14, 7-15
Black oil reservoir 4-10
Bonds
  coupon rate 2-9
  yield to maturity 2-9

BS&W content,oil prices 6-2
BTU content, gas prices 6-2

Capital budgeting 3-2
Capital recovery factor, time value of money 2-5
Capitalize 1-4
Carried interest
  deals 7-4
  Manahan, Herndon, and Abercrombie cases 11-3
Carry overs and carry backs 11-2
Cash flow diagram 1-3, 2-1
Cash flow model 1-2
Ceiling price, natural gas 6-5
Cessation of production clause, lease 7-3
Changing the form of cash flow stream 2-9
Classifications of Reserves 4-2 to 4-5
  Arps 4-2
  SEC 4-2
  SPE 4-4
Comparing alternatives 3-22 to 3-25
Compound interest 2-3
Condensate production, windfall profit tax 10-2
Constant GOR, secondary streams 5-40
Constant condensate yield, secondary streams 5-43
Continuous compounding, interest 2-7
Cost and revenue sharing, deals 7-6
Cost depletion 11-12, 11-13, 11-23
Costs
  economic limit 8-2
  investments 8-1
  operating 8-1
Cox, author 5-12, 5-13
Cumulative gas-oil ratio, reserves 4-10
Cumulative production equation 5-3, 5-8 to 5-10
Cutler, author 5-37, 5-38

Data fluctuations, decline curves 5-18
DCFROR, discounted cash flow rate-of-return 3-12, 3-14
DD+A
  defined 1-6
  tax versus financial 1-9
Deals
  complex example 7-11 to 7-21
  flowchart 7-9, 7-10
  technique to determine WI's and NRI's 7-6 to 7-21, 7-10
  tree 7-8, 7-9

Decline curves
  correcting for downtime 5-17, 5-19 to 5-26
  data fluctuations 5-18
  equation sheet 5-46
  exponential see exponential decline
  history 5-11
  hyperbolic see hyperbolic decline
  multiple wells on lease 5-17
  original oil-in-place 5-13
  overview 4-12 to 4-14
  permeability and skin factor 5-14
  plotting the data 5-16, 5-17
  rate-time equation 5-12
  shape factor 5-14
  theory 5-12
  use of loss ratio B-1 to B-3
Declining balance depreciation D-1, D-2
Degree of proof, reserves 4-2
Delay rental 11-5, 11-7
Depletion 1-4, 1-6, 11-5
  also see Federal income taxation
Depreciation 11-6, 11-10, 11-14
  accelerated cost recovery system (ACRS) 1-9
    table of ACRS rates 11-12, 11-13
  classes of property 11-10, 11-12
  declining-balance D-1, D-2
  defined 1-4, 1-6
  straight-line 1-6, D-1
  sum-of-year's digits D-4
  units-of-production 1-6, D-3
  also see Federal income taxation
Developed fields, reserves 4-8
Development status, reserves 4-2
Discount factor 2-4
Discounted-cash-flow rate-of-return 3-12, 3-14
Discounted profit-to-investment ratio 3-11,3-12
Discounting
  mid period 2-8
  partial first period 2-8
Domestic gas production eligible for percentage depletion 11-13
Drainage area, reserves 4-7

Earlougher, author 5-13
Economic interest concept, federal income taxation 11-2, 11-3, 11-6
Economic limit calculation 8-2
Effective decline rate 5-2, 5-3, 5-8
Effective interest rate 2-7

Ehlig-Economides, author 5-12, 5-14
Equation sheets
- compound interest equations 2-13
- decline curve equations 5-46

Equivalence, concept of 2-1
Estimated remaining reserves 4-1
Estimated ultimate recovery 4-1, 4-6
Existing interstate contracts, NGPA 6-6
Expenditure limits, operating agreements 7-6
Expense 1-4
Exponential decline
- cumulative production equation 5-3
- effective decline rate 5-2, 5-3
- nominal decline rate 5-2, 5-3
- rate-cum plots 5-18, 5-19
- rate-time equation 5-2
- rate-time plots 5-18

Farm out 7-4
- 1/3 for 1/4 7-4

Federal income taxation
- ACRS 11-10, 11-12, 11-16
- allocated overhead 11-7
- allowable deductions 11-2, 11-6
- carried interest 11-3
- carried interest, Manahan, Herndon, and Abercrombie cases 11-3
- carry overs and carry backs 11-2
- cost depletion 11-12, 11-13, 11-23
- delay rental 11-5, 11-7
- depletion allowance 11-5, 11-15, 11-16, 11-18
- depreciables 11-6, 11-10, 11-14
- depreciation
  - classes of property 11-10, 11-12
  - table of ACRS rates 11-12, 11-13
- detailed model 11-6 to 11-23
- domestic gas production eligible for percentage depletion 11-3
- economic interest concept 11-2, 11-3, 11-6
- geological and geophysical costs 11-5
- geological area of interest 11-5
- geological project costs 11-5
- gross income 11-1, 11-6
- intangible drilling and development costs (IDC's) 11-6, 11-7, 11-17, 11-20
  - election to expense 11-7, 11-10
  - table of example IDC's 11-9
- interest expenses 11-7
- investment tax credit (ITC) 11-16, 11-19, 11-22

IPRO
percentage depletion rates 11-16
production eligible for percentage depletion 11-13, 11-16
lease and well equipment 11-6
lease bonus 11-5
lease operating expenses 11-6, 11-7
leasehold costs 11-5
minimum tax 11-2, 11-20, 11-21
negative cash tax 11-1
net income limitation 11-16
net profits interest 11-3
overriding royalty interest 11-3
percentage depletion 11-13, 11-15, 11-16, 11-21
eligibility for 11-13
production payment 11-4
property concept 11-4, 11-5
separate deposit 11-4
separate interest 11-5
separate tract 11-4
royalty interest 11-3
simplified model 11-1
state and local taxes 11-7
tangibles 11-6, 11-9, 11-14, 11-16
tax credits 11-2
taxable income 11-2
working interest 11-3
Fetkovich type curves 5-19 to 5-33
Fetkovich, author 5-12, 5-19, 5-27, 5-39, 5-40
Financial model 1-4
Financial net income 1-4, 1-5, 1-6
Formation volume factors
gas 4-10
oil 4-7, 4-10
total 4-10
Future worth of lump sum 2-3

Garb, author 5-43
Gas oil equivalence 6-2, 6-3
Gas prices
BTU content 6-2
regulation
inflation adjustment factor 6-5
celling price 6-5
existing interstate contracts 6-6
history 6-4
marker well 6-5
new gas supplies 6-5, 6-6

        new onshore reservoir   6-6, 6-7
        new onshore well   6-5
        OCS Lease   6-6
        real growth factor   6-5
        section 102 gas   6-5, 6-7
        section 103 gas   6-6, 6-7
        section 104 gas   6-6
        section 105 gas   6-6
        section 106 gas   6-6
        section 107 gas   6-6, 6-7
        section 108 gas   6-6
        section 109 gas   6-6
        stripper well natural gas   6-7
        tight gas   6-6
    water content   6-2
Gas condensate reservoir   4-10
Gas formation volume factor   4-7, 4-10
Gentry, author   5-34
Geological and geophysical, federal income taxation   11-5
Geological area of interest   11-5
Geological project costs   11-5
Geometric gradient formula, derivation   A-1 to A-2
GNP implicit price deflator   10-3, 10-12
GOR vs Np plot   4-12
Graph paper, construction of specially scaled   C-1
Gross income, federal income taxation   11-1, 11-6

Harmonic decline   5-5
Heavy oil, windfall profit tax   10-2
Hyperbolic decline
    cumulative production equation   5-8 to 5-10
    Di from tangent   5-37
    effective decline rate   5-8
    Fetkovich type curves   5-19 to 5-33
    harmonic   5-5
    nominal decline rate   5-5
    rate-time equation   5-4
    special graph paper   5-37
    trial and error   5-33 to 5-36
    values of b   5-5 to 5-7, 5-9, 5-37 to 5-39
Hyperbolically declining yield, secondary streams 5-43, 5-44

Imputed cost depletion, windfall profit tax   10-5, 10-6, 10-7
Incremental analysis concept   1-11
Incremental tertiary oil, windfall profit tax   10-2

Independent producer and royalty owner (IPRO)
    percentage depletion rates 11-16
    production eligible for percentage depletion 11-13, 11-15
    windfall profit tax 10-8
Inflation adjustment factor
    calculation, windfall profit tax 10-3
    table of 10-12
Intangible drilling and development costs 11-6, 11-7, 11-17, 11-20
    election to expense 11-7, 11-10
Interest
    compound 2-3
    concept 2-1
    continuous compounding 2-7
    effective 2-7
    nominal 2-7
    simple 2-2
Investment decisions
    capital budgeting 3-2
    mutually exclusive projects 3-1
    single project 3-1
Investment yardsticks
    accounting rate-of-return 3-3, 3-5
    appreciation-of-equity rate-of-return 3-21
    attributes of ideal 3-2
    bailout factor 3-3, 3-4
    discounted-cash-flow rate-of-return 3-12, 3-14
    discounted profit-to-investment ratio 3-11, 3-12
    modified DCFROR 3-16, 3-20
    net present value 3-5, 3-7
    payback 3-2
    present value profile 3-8, 3-13
    undiscounted profit-to-investment ratio 3-5, 3-6
    urgency 3-2
Investment tax credit 11-16, 11-19, 11-22
Isolated wells, reserves 4-8
Isovol map, reserves 4-9

Land and legal
    net revenue interest 7-7, 7-8
    working interest 7-7, 7-8
Lease and well equipment, federal income taxation 11-6
Lease bonus, federal income taxation 11-5
Lease operating expenses, federal income taxation 11-6, 11-7
Lease revenue interest 7-9
Leasehold costs, federal income taxation 11-5
Leasehold interest

cessation-of-production clause 7-3
deals 7-1
lease bonus 7-3
pooling/unitization clause 7-3
royalty interest 7-3
shut-in gas well clause 7-3
Lefkovits, author 5-11
Lewis, author 5-1
Log Gp vs log Np plot, secondary streams 5-40, 5-41
Log Np vs Log Gp plot, secondary streams 5-45

Marker well 6-5
Matching principle 1-4
Material balance, reserves 4-9 to 4-11
Matthews, author 5-11
Mineral interest 7-1
also see oil and gas rights
Minimum tax 11-2, 11-20, 11-21
Modified DCFROR 3-16, 3-20
Modified sinking fund 2-6
Multiple rates-of-return 3-15
Multiple wells on lease, reserves 5-17
Mutually exclusive projects 3-1

N vs Np plot 4-10
National petroleum reserves, windfall profit tax 10-2
Negative cash tax 11-1
Net cash flow
defined 1-2
for oil property 1-13
Net income limitation
federal income taxation 11-16
windfall profit tax 10-1, 10-4, 10-5, 10-8
Net present value 3-5, 3-7
Net profits interest
deals 7-5
federal income taxation 11-3
Net revenue interest
deals 7-1, 7-2, 7-6
how to determine 7-6 to 7-21
Net revenue interest
land and legal 7-7, 7-8
oil property evaluation 7-2, 7-8
New gas supplies, price regulation 6-5, 6-6
New onshore reservoir, price regulation 6-6, 6-7
New onshore well, price regulation 6-5

Newly discovered oil, Windfall profit tax 10-2
Nominal decline rate 5-2, 5-3, 5-5
Nominal interest rate 2-7
Non cash charges, see DD$A
Non-consent clause, operating agreements 7-6
Non-operating interest, defined 7-5

OCS Lease price regulation 6-6
Oil prices
  API gravity 6-1
  BS&W content 6-2
  price bulletins 6-2
  regulation 6-3, 6-4
  shrinkage 6-2
  specific gravity 6-1
Oil and gas rights 7-1 to 7-5
  carried interest 7-4
  leasehold interest 7-1
  mineral interest 7-1
  net profits interest 7-5
  overriding royalty interest 7-4
  primary term 7-3
  production payment 7-5
  reversionary interests 7-4
  royalty interest 7-3
Oil formation volume factor 4-7, 4-10
Oil property evaluation
  net revenue interest 7-2, 7-8
  working interest 7-2, 7-8
Operating agreements 7-5 to 7-6
  accounting procedure 7-6
  COPAS 7-6
  cost and revenue sharing 7-6
  expenditure limits 7-6
  non-consent clause 7-6
  operator responsibilities 7-6
Operating costs 8-2
Operating interest , defined 7-5
Original oil-in-place, decline curves 5-13
Overriding royalty interest
  deals 7-4
  federal income taxation 11-3
  also see oil and gas rights and Federal income taxation

P/Z vs Gp plot
  effect of rock compressibility 4-11, 4-12
  volumetric reservoir 4-11, 4-12
  water drive 4-11, 4-12
Payback 3-2
  also see investment yardsticks
Percentage depletion
  eligibility 11-13
  federal income taxation 11-13, 11-21
Periodicity 1-1
Permeability and skin factor 5-14
Plotting the data, decline curves 5-16, 5-17
Pooling/unitization clause, leases 7-3
Porosity 4-7
  average 4-8
Present value profile 3-8, 3-13
Present worth of lump sum 2-3
Pressure vs Np plot 4-12
Price bulletins, oil prices 6-2
Primary term, lease 7-3
Producing status, reserves 4-2
Production payment
  deals 7-5
  federal income taxation 11-4
Profit
  defined 1-1
  financial model 1-4
  model used for investment analysis 1-10
  net cash flow 1-2
  tax model 1-4
Property concept, federal income taxation 11-4, 11-5
Proportionately reduced, deals 7-4
Proved developed reserves, defined 4-4, 4-5
Proved reserves, defined 4-2, 4-4
Proved undeveloped reserves, defined 4-4, 4-5

Ramsay, author 5-37, 5-38
Rate-time equations, decline curves 5-2, 5-4, 5-12, 5-18, 5-46
Rate-cum plots 5-18, 5-19
Real growth factor, price regulation 6-5
Recovery factor
  analogy 4-6
  defined 4-1
  volumetrics 4-7
Regulation history, gas prices 6-4
Regulation
  annual inflation adjustment factor 6-5

Report content
  appendix 12-3
  conclusions 12-1
  discussion 12-2
  introduction 12-1
  recommendations 12-2
Reserves
  allocating to individual wells 4-9
  analogy 4-6, 4-13
  barrels per acre-foot 4-6
  basic terminology 4-1, 4-2
  classification 4-2 to 4-5
    Arps 4-2
    Arps table 4-3
    authors editorial 4-4 to 4-6
    degree of proof 4-2
    development status 4-2
    producing status 4-2
    SEC 4-2
      proved developed 4-4, 4-5
      proved 4-2, 4-4
      proved undeveloped 4-4, 4-5
    source of energy 4-2
    SPE 4-4
  cumulative recovery 4-1
  decline curve, overview 4-12 to 4-14
  developed fields
  drainage area 4-7
  effect of rock compressibility 4-11, 4-12
  effective date 4-1
  estimated remaining reserves 4-1
  estimated ultimate recovery 4-1, 4-6
  gas formation volume factor 4-7
  isolated wells 4-8
  material balance 4-9 to 4-11
  oil formation volume factor 4-7, 4-10
  original 4-1
  porosity 4-7
  recovery factor 4-1, 4-6, 4-7
  reservoir simulation 4-12 to 4-15
  summary table for methods to calculate 4-13 to 4-15
  volumetric 4-7, 4-13
  volumetric, developed fields 4-8
  water saturation 4-7
Reservoir simulation 4-12 to 4-15
Reversionary interest
  see Oil and gas rights
Reversions 7-4

how to handle cost and revenue sharing 7-11
Royalty interest
deals 7-3
federal income taxation 11-3
Royalty owner exemption, windfall profit tax 10-8, 10-9 ,10-11
Rule of 72 2-6
Rule of 78 2-11

Secondary streams
forecasting condenstate
constant yield 5-43
hyperbolically declining yield 5-43, 5-44
log Np vs Log Gp 5-45
forecasting gas
constant GOR 5-40
log GP vs log Np 5-40, 5-41
forecasting water, WOR vs Cum Oil 5-41, 5-43
Section 102 gas 6-5, 6-7
Section 103 gas 6-6, 6-7
Section 104 gas 6-6
Section 105 gas 6-6
Section 106 gas 6-6
Section 107 gas 6-6, 6-7
Section 108 gas 6-6
Section 109 gas 6-6
Separate deposit, Federal income taxation 11-4
Separate interest, Federal income taxation 11-5
Separate tract, Federal income taxation 11-4
Severance tax adjustment, windfall profit tax 10-1, 10-3, 10-4
Severance taxes 9-1 to 9-5
Shape factor 5-14
Shrinkage 6-2
Shut-in gas well clause, leases 7-3
Simple interest 2-2
Simplified model, Federal income taxation 11-1
Single project, Investment decisions 3-1
Sinking fund 2-4
factor 2-4
uniform periodic series 2-4
Solution gas-oil ratio 4-10
Source of energy, reserves 4-2
Specific gravity 6-1
State and local taxes 11-7
Straight line depreciation 1-6, D-1
Stripper oil, windfall profit tax
base price 10-2
exemption 10-11

Stripper well natural gas, gas price 6-7
Sum-of-year's digits depreciation D-4

Table of WI's and NRI's, Technique to determine 7-10
Tangibles, federal income taxation 11-6, 11-9, 11-14, 11-16
Tax credits, federal income taxation 11-2
Tax rates, windfall profit tax 10-8, 10-9
Taxable income, defined 11-2
Tier 1 oil 10-2
Tier 2 oil 10-2, 10-3
Tier 3 oil 10-2, 10-3
Tight gas, gas prices 6-6
Time 0 concept 1-2
Time value of money
  annuity (loan) 2-5
  annuity due or lease 2-5
  capital recovery factor 2-5
  changing the form of cash flow stream 2-9
  discount factor 2-4
  future worth of lump sum 2-3
  modified sinking fund 2-6
  present worth of lump sum 2-3
  sinking fund 2-4
  sinking fund factor 2-4
Transfer rules, windfall profit tax 10-11
Trigger points
  after casing point and before payout 7-11, 7-15
  after payout 7-11, 7-15, 7-16
  before casing point 7-11, 7-13, 7-14, 7-15
  deals 7-11

Uncertainty 1-13
Undiscounted profit-to-investment ratio 3-5,3-6
Uniform annual annuity 2-9
Units-of-production depreciation 1-6, D-3

Values of 'b' 5-5 to 5-7, 5-9, 5-37 to 5-39
Volatile oil reservoir 4-10
Volumetric reserves 4-7, 4-13

WI's and NRI's, Technique to determine 7-6 to 7-21
WOR vs Cum oil 5-41, 5-43
Water content, gas prices 6-2
Water saturation, average 4-8

Windfall profit tax
- adjusted base price 10-1, 10-2, 10-3
- base price 10-2
- condensate production 10-2
- exemption for charitable organizations 10-11
- exemption for royalty owners 10-8, 10-9 ,10-11
- exemption for state and local governments 10-11
- GNP implicit price deflator 10-3, 10-12
- heavy oil 10-2
- imputed cost depletion 10-5, 10-6, 10-7
- incremental tertiary oil 10-2
- independent producer 10-8
- inflation adjustment factor calculation 10-3
- inflation adjustment factor table 10-12
- national petroleum reserve 10-2
- net income limitation 10-1, 10-4, 10-5, 10-8
- newly discovered 10-2
- phase out 10-1
- removal price 10-1, 10-2
- severance tax adjustment 10-1, 10-3, 10-4
- stripper oil classification 10-2
- stripper oil exemption 10-11
- tax rates 10-8, 10-9
- Tier 1 oil 10-2
- Tier 2 oil 10-2, 10-3
- Tier 3 oil 10-2, 10-3
- transfer rules 10-11

Working interest
- deals 7-1, 7-2, 7-6
- federal income taxation 11-3
- how to determine 7-6 to 7-21
- land and legal 7-7, 7-8
- oil property evaluation 7-2, 7-8